Validity and Reliability in Built Environment Research

"This provides a welcome addition to the literature on research methods in the built environment and should be useful for students at all levels".

Dr. Mark Addis
London School of Economics and Political Science

This book aims to guide researchers who are engaged in social science and built environment research through the process of testing the reliability and validity of their research outputs following the application of different methods of data collection.

The book presents case studies that emphasize reliability and validity in different examples of qualitative, quantitative and mixed method data sets, as well as covering action research and grounded theory. The reader is guided through case studies that demonstrate:

- An understanding of the reliability and validity approaches from social science and built environment perspectives in alignment with the relevant research philosophies, approaches and data collection strategies
- Real research projects that have been conducted by expert researchers on topics such as Lean, BIM, Housing and Sustainability to answer specific or evolving questions in relation to the reliability and validity of research
- A simple and easy method that students at Masters and PhD levels can relate to in order to adopt a sound reliability and validity approach to their research

This book is the essential guide for researchers at undergraduate and postgraduate level who need to understand how to validate the quality of the empirical tests they conduct using different techniques. The book will also be a great asset to supervisors from different backgrounds who need a refresher on this key aspect of the research cycle.

Vian Ahmed has over 25 years of industrial and academic experience in the United Kingdom and overseas. During her employment at the University of Salford (2004–2018), she took on a number of management positions, and became

Professor in the Built Environment in 2010. She was the Director of the Online Doctoral Programme at the School of the Built Environment (2004–2016) and the Director of Postgraduate Research (2007–2016). Her previous books include: *Research Methodology in the Built Environment* (2016), and *Leadership and Sustainability in the Built Environment* (2015).

Alex Opoku is an Associate Professor in Quantity Surveying & Construction Management at the College of Engineering, University of Sharjah, UAE. He is a Fellow of the UK Higher Education Academy (FHEA), Chartered Quantity Surveyor (MRICS) and Chartered Construction Manager (MCIOB) with many years of teaching and learning experience in the UK Higher Education sector and the UK construction industry.

Ayokunle Olubunmi Olanipekun is a Postdoctoral Research Fellow in Construction Management in the School of Built Environment at Massey University, New Zealand. He has professional academic experience in higher education institutions in Nigeria, Australia and New Zealand since 2011. He is Fellow of the UK Higher Education Academy and a Professional Member of the Nigerian Institute of Quantity Surveyors (NIQS).

Monty Sutrisna is Professor of Construction and Project Management and the Head of School – School of Built Environment at Massey University, New Zealand. With professional experience in both industry and academia in the United Kingdom, Australia, New Zealand and Indonesia for about 20 years, he has been championing close collaboration between industry and academia to achieve synergy. He is a Fellow of the UK Higher Education Academy, a Fellow of the Australian Institute of Building, a fellow of the Royal Institution of Chartered Surveyors and also a member of the Institution of Civil Engineers, Chartered Institute of Building and Chartered Institution of Civil Engineering Surveyors.

Validity and Reliability in Built Environment Research

A Selection of Case Studies

Edited by
Vian Ahmed, Alex Opoku,
Ayokunle Olanipekun
and Monty Sutrisna

Routledge
Taylor & Francis Group

LONDON AND NEW YORK

First published 2022
by Routledge
2 Park Square, Milton Park, Abingdon, Oxon OX14 4RN

and by Routledge
605 Third Avenue, New York, NY 10158

Routledge is an imprint of the Taylor & Francis Group, an informa business

British Library Cataloguing-in-Publication Data
A catalogue record for this book is available from the British Library

Library of Congress Cataloging-in-Publication Data
Names: Ahmed, Vian, editor. | Opoku, Alex, editor. | Olanipekun, Ayokunle, editor. | Sutrisna, Monty, editor.
Title: Validity and reliability in built environment research: a selection of case studies/edited by Vian Ahmed, Alex Opoku, Ayokunle Olanipekun and Monty Sutrisna.
Description: Milton Park, Abingdon, Oxon; New York, NY: Routledge, 2022. | Includes bibliographical references and index.
Identifiers: LCCN 2021045703 (print) | LCCN 2021045704 (ebook) | ISBN 9780367197766 (hbk) | ISBN 9780367197803 (pbk) | ISBN 9780429243226 (ebk) | ISBN 9780429512940 (adobe pdf) | ISBN 9780429519802 (mobi) | ISBN 9780429516375 (epub)
Subjects: LCSH: Reliability (Engineering)–Case studies. | Testing–Case studies. | Building–Research–Case studies. | Social sciences–Research–Case studies. | Business–Research–Case studies.
Classification: LCC TA169 .V34 2022 (print) | LCC TA169 (ebook) | DDC 620/.00452–dc23
LC record available at https://lccn.loc.gov/2021045703
LC ebook record available at https://lccn.loc.gov/2021045704

ISBN: 978-0-367-19776-6 (hbk)
ISBN: 978-0-367-19780-3 (pbk)
ISBN: 978-0-429-24322-6 (ebk)

DOI: 10.1201/9780429243226

Typeset in Goudy
by KnowledgeWorks Global Ltd.

Contents

List of illustrations

Figures

Contributors

Editors

Vian Ahmed is a Senior Fellow of the UK Higher Education and a Fellow of the Chartered Institute of Building. She has gathered over 25 years of industrial and academic experience in the United Kingdom and overseas. She is currently a Professor in the Industrial Engineering Department at the American University of Sharjah and the Director of Teaching and Learning Alternative Delivery for the College of Engineering. She obtained a (BEng.) in Civil Engineering, (MSc) and (PhD) in Construction. She has broad expertise in teaching at undergraduate and postgraduate levels, with expertise in construction project management and IT. She has over 30 graduated PhD students within the Built Environment discipline, covering different research themes such as sustainability, energy saving, intelligent designs, Building Information Modelling, PPP/PFI, e-learning, disaster and resilience management. She has chaired and organized a number of national and international workshops and seminars and secured a number of research grants, with more than 100 refereed journal and conference papers.

Alex Opoku is an Associate Professor in Quantity Surveying & Construction Management at the University of Sharjah, UAE. He is a Fellow of the UK Higher Education Academy (FHEA), Chartered Quantity Surveyor (MRICS) and Chartered Construction Manager (MCIOB) with many years of teaching and learning experience in the UK Higher Education sector and the UK construction industry.

Ayokunle Olubunmi Olanipekun is a Senior Lecturer in Quantity Surveying in the School of Architecture and Built Environment, University of Wolverhampton, UK. He has professional academic experience in higher education institutions in Nigeria, Australia and New Zealand since 2011. He is Fellow of the UK Higher Education Academy and a Professional Member of the Nigerian Institute of Quantity Surveyors (NIQS)

Monty Sutrisna is Professor of Construction and Project Management and the Head of School – School of Built Environment at Massey University, New Zealand. With professional experience in both industry and academia in the

United Kingdom, Australia, New Zealand and Indonesia for about 20 years, he has been championing close collaboration between industry and academia to achieve synergy. He is a Fellow of the UK Higher Education Academy, a Fellow of the Australian Institute of Building, a fellow of the Royal Institution of Chartered Surveyors and also a member of the Institution of Civil Engineers, Chartered Institute of Building and Chartered Institution of Civil Engineering Surveyors.

Authors

Ahmad Saad is an Engineering System Management MSc graduate from the College of Engineering at the American University of Sharjah, with a Bachelor Degree in Electrical Engineering. He is currently working as a Data Engineering Analyst in one of the major consulting companies in the UAE.

Alia Al Sadawi is a doctoral candidate doing her PhD in Engineering System Management. She has completed a bachelor's degree in Electronics Engineering from Ajman University and Master degree in Engineering Systems Management from The American University of Sharjah in 2016. Her research interests include engineering management, smart cities management and advanced decision-making analysis.

Ayman Alzaatreh is an Associate Professor of Statistics at the American University of Sharjah. He has a PhD in Statistics, BSc and MSc in Mathematics. He has more than 12 years of teaching experience in the United States and overseas. His current research interest focuses on distribution theory, statistical inference of probability models, multivariate weighted distribution and data mining. He has more than 50 referred journal and conference publications and chaired and organized number of international conferences and workshops.

Ayomikun Solomon Adewumi completed a doctorate in Architecture and Urban planning at University of Dundee, United Kingdom in July 2020. The thesis explored how urban sustainability can be delivered at the neighbourhood scale of spatial planning through the adoption of indicators. Prior to this, he had both Bachelor of Technology (B.Tech) and Master of Technology (M.Tech) degrees in Architecture at the Federal University of Technology Akure, Nigeria. To date, he has published 3 journal articles and presented a conferences and workshops. He is a Fellow of the Higher Education Academy (FHEA). Currently, he lectures at London South Bank University, United Kingdom.

Tahani Alnaqbi is postgraduate researcher at Engineering System Management in the College of Engineering at the American University of Sharjah and conducting a research in innovative solutions regarding the water scarcity issues in the UAE region. She is also working with the Government of Sharjah as the Head Manager of the Building Permits Section since 2015. In 2018, she was awarded Honorary Best Leading Manager in the organization.

Her future aspirations include working as an Engineer who introducing artificial intelligence to develop the urban development in the UAE.

Nikita Kasianov is an Engineering System Management MSc graduate from the College of Engineering at the American University of Sharjah, with a Bachelor Degree in Civil Engineering. Upon graduation, Nikita worked as a site engineer on a number of projects within the construction industry including the construction of a concrete laboratory and recreation facilities.

Heba Khlaif is an Engineering System Management MSc student in the College of Engineering at the American University of Sharjah, with a Bachelor Degree in Industrial Engineering and prior experience in project management & cargo commercial product development.

Dumiso Moyo is a Chartered Town Planner (MRTPI) and a Senior Fellow of the Higher Education Academy (SFHEA). He is author of the book *Explaining the Low-income Housing Dilemma in Sub-Saharan Africa*. Currently, he is a Lecturer at the University of Dundee, United Kingdom, and researches on cities in sub-Saharan Africa.

Malick Ndiaye is an Associate Professor in the American University of Sharjah. He has a PhD in Operations Research from the University of Bourgogne, France. He has worked for King Fahd University of Petroleum and Minerals (KFUPM) and the Management Mathematics Program at the University of Birmingham, United Kingdom. His research areas cover operations research, supply chain management and location theory and its applications to GIS. His research has received grants funded by the Capital Region of Brussels, Belgium; the University of Birmingham; and KFUPM. He is a Certified Supply Chain Professional from the American Association for Operations Management (APICS) and a qualified APICS trainer.

Edoghogho Ogbeifun is a Senior Lecturer in the Department of Civil Engineering, University of Jos and Research Fellow in the Postgraduate School of Engineering Management, University of Johannesburg. He has over 35 years work experience in the built environment development. He is a Professional civil engineer, Council for the Regulation of Engineering in Nigeria (COREN) and an accredited Facilities Professional (AFP) of the South African Facilities Management Association.

Vincent Onyango is a Lecturer and researcher at the University of Dundee's Department of Architecture and Urban Planning and leads the degree programme Environmental Sustainability. His research focuses on the interface between the natural and built environments, covering environmental assessments, integrated environmental planning and management, with greater interest in processes and methodologies towards unifying the broader and more integrative issues of sustainability and environmental governance. He has recently undertaken research on the effectiveness and design of

Scotland's policies on greenhouse gas emissions in relation to new houses, marine planning, and implications for multi-use of the oceans.

Jan-Harm C. Pretorius is a trained Baldrige (USA) and South African Excellence Foundation (SAEF) assessor. He worked at the South African Atomic Energy Corporation (AEC) as a Senior Consulting Engineer for 15 years. He is currently a Professor and Head of School: Postgraduate School of Engineering Management in the Faculty of Engineering and the Built Environment, University of Johannesburg.

Sara Saboor is a Doctoral Candidate in Engineering System Management program at the American University of Sharjah. She has an engineering background with Bachelor's in Electrical Engineering (Electronics) and Master's degree in Electrical Engineering (Telecommunication) from the National University of Sciences and Technology (NUST), one of the top universities in Pakistan and recognized worldwide. She works as a Graduate teaching/research assistant at the American University of Sharjah. Her research interests include engineering management, strategic management, HR management and advanced decision-making analysis.

Hasan Saleh is a graduate student in the Engineering Systems Management program at the American University of Sharjah. He holds a bachelor degree in civil engineering since 2016 and has started his career in the construction industry upon graduation. He shifted to facilities management in 2017 and has been working to utilize his knowledge and skills in this field.

Willie Tan is tenured Professor, Department of the Built Environment, College of Design and Engineering, National University of Singapore. He specializes in geomatics, urban management and project finance. He is Editor of the World Scientific Series on the Built Environment and an editor of Spatial Science. He has written books on research methods and project finance.

Dana Yazbak is an Engineering System Management MSc student in the College of Engineering at the American University of Sharjah, with a Bachelor Degree in Civil Engineering and prior experience in Demand & Supply Planning with the Cosmetic Supply Chain Industry.

Yajian Zhang worked as a Research Fellow in National University of Singapore from 2017 to 2019. His research interests include managing large infrastructure projects and sustainable urban development. Currently, he is pursing another graduate degree in information systems.

Acknowledgements

The editors have been part of many undergraduate and postgraduate research journeys over the years as educators, advisors, investigators and reviewers, and have taken part in a number of academic debates in relation to the validity and reliability of research, which motivated them to produce this book. In return, the editors would like to acknowledge the contribution of the academic community within the engineering and built environment discipline at large for never ceasing to innovate, challenge existing practices and create new knowledge for the betterment of this world.

As editors, we would like to thank all our colleagues and students who have contributed to our research journey and topped up our realization of the importance of the validity and reliability of research outputs. We also thank all the authors who have contributed chapters that made this book a reality, and for their patience while surviving the challenges that COVID-19 imposed upon them and us. We thank our families who supported us while taking time out to produce this book.

Last but not the least, we would also like to thank Routledge publishers for helping us realize our long-term ambition to produce this book, and understanding the existing need for sharing examples of some of the good practices for the validity and reliability of research in engineering and the built environment.

Vian Ahmed
Alex Opoku
Ayokunle Olanipekun
Monty Sutrisna

Introduction

This book aims to guide researchers who are engaged in social science and built environment research through the thought process of testing the reliability and validity of the research outputs following the application of different methods of data collection. This book is very much inspired by our recently published book on Research Methodology in the Built Environment: A Selection of Case Studies (2016), which gathered a number of case studies that illustrate the thought process of applying different research methods by using examples of qualitative, quantitative and mixed method research as well as action research and grounded theory. However, the book did not look into the final stages of the research, hence the reliability and validity of the results, which is of great importance to any research. Therefore, to achieve the intended aim of this book, the reader will be guided through show-casing quality research that demonstrates;

- A simplified understanding of the reliability and validity approaches from the social science and built environment prospective in alignment with the relevant research philosophies, approaches and data collection strategies. The book will be structured in a form of a selection of case studies that bring together a comprehensive range of different scenarios that cover various data collection strategies such as; qualitative, quantitative, mixed methods research etc.
- An overview of different case scenarios that have been formed by researchers within social science and built environment disciplines, to answer specific or evolving questions in relation to the reliability and validity of research in a simple and easy way that students at Masters and PhD levels can relate to. The book will address the fundamental issues that researchers must identify in order to adapt a sound reliability and validity approach.

Validity and Reliability are integral parts of any social science research in order to judge the quality of its research design and to test the quality of any empirical tests that have been adapted. The authors have observed a large number of dissertation and thesis at Postgraduate and Undergraduate levels that pay little attention to this aspect of the study, raising many questions about the integrity of the research. The authors have also observed that a large number of PhD thesis are being resubmitted in order to complete the validity and reliability aspect of the research, which is also often the focus of many debates during a PhD viva. This book will be a great asset for researchers who are perusing their research at UG and PG research to understand the full cycle of a research journey with a particular focus on reliability and validity. The book will also be a great asset to supervisors who come from different education backgrounds and are not necessarily well informed of this aspect of the research.

Although there is a broad range of resources that explain the theoretical aspects of validity and reliability in Medical, Engineering and Social Science research, there is no one resource that demonstrates this concept by covering the thought process of the research as a whole, with the defined pivotal aspects for various methodologies and data collection methods in order to demonstrate the importance validity and reliability to confirming the quality of the research. This approach will provoke researchers to how to think of their research, and where reliability and validity comes in within that journey.

The book consists of ten chapters split into the following four parts:

Part I includes two chapters that provide the reader with a good understanding of the thought process of research with all the defined pivots, with a particular focus on the theoretical aspects of validity and reliability tests and how these can be aligned to the thought process.

Parts II, III and IV include a selection of case studies that demonstrate the validity and reliability tests and approaches used by researchers to affirm the quality of the research adapting different methodologies and data collection strategies.

Part I

Research reliability and validity

1 Understanding reliability in research

Vian Ahmed, Ayokunle Olanipekun, Alex Opoku and Monty Sutrisna

1.1 Measuring errors in research

An important part of Built Environment (BE) research is the measurement of social phenomena. The idea of measurement conforms with the positivist research paradigm, or the empirical analytic approach for discerning reality and explaining social phenomena in the process (Dainty, 2008; Drost, 2011). In line with the paradigm, measurement requires precise definitions of (social) conceptual meanings, and such concepts in the BE include safety, performance, in addition to design and social science concepts that are applicable to the functioning of the BE such as culture and motivation. These concepts are abstract concepts, which are only theoretically constructed (Kimberlin & Winterstein, 2008). Measurement involves the operationalization of the concepts into definite variables (or specific questions/items) and the application of measuring instruments (and scientific tests) to quantify them (Kimberlin & Winterstein, 2008). For example, successful project performance may be operationalized as "project completion within or under-budget" while the related measurement instrument may ascertain "data on cost information" of completed projects. A questionnaire (which may either structured or unstructured) is the most common measuring instrument for operationalizing abstract concepts in BE and consequently used to obtain data.

However, measurement errors are plausible when abstract concepts are operationalized and developed into questionnaires to obtain data in BE research. Compared with objective sources of data like laboratory test results, data obtained on abstract concepts using questionnaires involve greater subjectivity in judgement that leads to error in measurement (Kimberlin & Winterstein, 2008). Accordingly, Drost (2011) identifies two sources of measurement errors in research: systematic errors and random errors. For example, consider a bathroom scale, a systematic error is when the scale produces a consistent measure of a person's weight but was always 5lb higher than it should be. A random error is when the scale produces correct weight measure, but the person misread the weight value. Based on the classical test theory, measurement error is the difference between the true value and the measured value obtained by a measuring instrument (i.e. a questionnaire) (Kimberlin & Winterstein, 2008; Mohajan, 2017). In

DOI: 10.1201/9780429243226-2

a standard BE research, the value obtained by using a questionnaire is the sum of both the "true value" which is naturally unknown, and the "error" in the measurement process (Mohajan, 2017). Essentially, the "true value" is obtained if a measurement is perfectly accurate (Kimberlin & Winterstein, 2008). Therefore, pertaining to the measurement of abstract concepts, minimizing measurement errors to the barest minimum by enhancing the reliability of the measurement instrument and validity of the measured concepts will increase the rigour of BE research on social phenomena (e.g. Bannigan & Watson, 2009).

1.2 Research reliability

In quantitative BE research, measurements of social concepts are carried out by using measuring instruments (i.e. questionnaire). The measuring instrument is reliable when it yields consistently the same or comparable results over repeated measures (Drost, 2011). That is, regardless of who performs the measurement, and the occasion and condition under which measurement was carried out, the results produced by the measuring instrument is consistent (or comparably consistent) (Brink, 1993; Mohajan, 2017). Therefore, reliability can be regarded as the accuracy of a measuring instrument in quantitative BE research (Heale & Twycross, 2015). Take the bathroom scale example, if it consistently produces a person's true weight (e.g. 35lb) over repeated times, then it is a reliable measuring instrument. Therefore, for the BE researcher, the challenge of reliability is to develop measuring instruments to obtain the true values of measured concepts to reduce error in measurement process. This requires the testing of reliability of measuring instruments (Heale & Twycross, 2015). The three attributes of reliability that are often tested are: *stability, homogeneity or internal consistency* and *equivalence* as outlined in Table 1.1.

1.2.1 Stability

Stability refers to the ability of a measure to remain the same over time without controlling the testing conditions or respondent themselves (Mohajan, 2017). Therefore, a perfectly stable measuring instrument will produce the same results when administered time after time to collect data (Bannigan & Watson, 2009) and this is obtained by performing the test-retest reliability method.

Table 1.1 Attributes of reliability test

Attributes	Description
Stability	The consistency of results using an instrument with repeated testing
Homogeneity (or internal consistency)	The extent to which all the items on a scale measure one construct
Equivalence	Consistency among responses of multiple users of an instrument, or among alternate forms of an instrument

Source: Heale et al. (2015)

1.2.1.1 *Test-retest reliability method*

The test-retest reliability refers to the temporal stability of test from one measurement session to another (Drost, 2011). It is obtained by administering the same test twice, or more over a period ranging from few weeks to months, on a group of individuals (respondents) (Mohajan, 2017) under similar circumstances (Heale & Twycross, 2015). The procedure is to administer the test to a group of respondents and then administer the same test to the same respondents later (Drost, 2011). Thereafter, a statistical comparison is made between participant's test scores (values) for each of the times they have completed it to provide an indication of the reliability of the instrument (Heale & Twycross, 2015). For example, construction workers may be asked to complete the same questionnaire about safety satisfaction twice in three months so that test results can be compared to assess stability of scores. The correlation coefficient calculated between two sets of data, and the higher the coefficient, the better the test-retest reliability (and stability).

(Mohajan, 2017)

Qu et al. (2009) studied the quality of life of migrant construction workers in Shenyang, China by using the 36-Item Short Form Health Survey (SF-36) questionnaire. The questionnaire is divided into eight domains of individual questions about the physical function, role limitations due to physical problems, bodily pain, general health, vitality, social functioning, role limitations due to emotional problems and mental health. The study was designed to evaluate the quality of life of the migrant construction workers at one-week time apart and therefore, the test-retest reliability method was performed to demonstrate the stability of the questionnaire over time. In the first time, a total of 1125 SF-36 questionnaires were administered to the migrant construction workers, and a week a later, the questionnaires were administered to 50 of them who were randomly selected. The correlation test was used to perform statistical comparison between the migrant workers' responses to the questionnaires in the first time and the week later. As shown in Table 1.2, the retest of the correlation between the items showed that $r > 0.70$ could be achieved for all eight domains ($P < 0.01$) (Table 1.2), demonstrating relatively good stability for the SF-36 questionnaire. The high correlation values indicate that the responses of the migrant construction workers about their quality of life remained uniform/consistent despite responding at different times, and a low correlation value would suggest otherwise. According to Ajayi (2017), the high correlation signifies high reliability of the SF-36 questionnaire administered to the migrant construction workers.

1.2.1.2 *Limitation*

Test-retest reliability is defined by the correlation between scores (values) on the identical tests given at different times (Drost, 2011) and this leads to some limitations. For instance, when the interval between the first and second test

Table 1.2 Test-retest reliability results

Domains	Test-retest reliability results (n = 50)
Physical function (PF)	0.801**
Role limitation due to physical problems (RP)	0.781**
Bodily pain (BP)	0.856**
General health (GH)	0.721**
Vitality (VT)	0.962**
Social functioning (SF)	0.841**
Role limitation due to emotional problems (RE)	0.78
Mental health (MH)	0.793**

**$P < 0.01$

Source: Qu et al. (2009)

is too short, respondents might remember what was on the first test and their answers on the second test could be affected by memory. Alternatively, when the interval between the two tests is too long, maturation happens – which is the changes in the subject factors (measured variables) or respondents that occur over time and cause a change from the initial measurements to the later (Drost, 2011). During the time between the two tests, the respondents could have been exposed to things which changed their opinions, feelings or attitudes about the behaviour under study (Drost, 2011; Thanasegaran, 2009). Ideally, the interval between administrations should be long enough that values obtained from the second administration will not be affected by the previous measurement but not so distant that learning or a change in health status could alter the way subjects respond during the second administration.

(Kimberlin & Winterstein, 2008)

1.2.2 Internal consistency

Internal consistency (or homogeneity) concerns the reliability within the measuring instrument and it questions how well a set of items (or variables) measures a concept that is being tested (or measured) (Drost, 2011). According to Kimberlin & et al. (2008), the assumption of internal consistency is that items (or variables) measuring the same concept should correlate, and therefore, the coefficient of internal consistency provides an estimate of the reliability of measurement. In other words, the more interrelated (unidimensional) the items are, the higher the calculated reliability coefficient (estimate) (Ekolu & Quainoo, 2019). The estimate is obtained by calculating the average intercorrelations among all single items (or variables) in a concept, or a test ((Drost, 2011) using one or more of the following methods: split-half reliability, Kuder-Richardson coefficient, Cronbach's alpha and inter-item consistency (inter-rater reliability) (Thanasegaran, 2009). However, there is no clarity around the interpretation of reliability estimates but estimates < 0.5 have been considered acceptable in research (Ekolu & Quainoo, 2019).

1.2.2.1 Split-half

The split-half method measures the degree of internal consistency by checking one half of the results of a set of scaled items in a measuring instrument against the other half (Thanasegaran, 2009). It requires only one administration of the measuring instrument (Mohajan, 2017), but the items in the instrument are split in half in several ways, for example, first half and second half, or by odd and even numbered items, to form two new measures testing the same social phenomena (Drost, 2011). In contrast to the test-retest reliability, the split-half method is usually measured in the same time (Drost, 2011). When the results are divided into in half, correlations are calculated comparing both halves (Heale & Twycross, 2015). Indeed, strong correlations indicate high reliability, while weak correlations indicate the instrument may not be reliable (Drost, 2011; Heale & Twycross, 2015). The method demands equal item representation across the two halves of the instrument, otherwise, the comparison of dissimilar sample items will not yield an accurate reliability estimate.

(Drost, 2011; Thanasegaran, 2009)

The reliability of the 36-Item Short Form Health Survey (SF-36) questionnaire for investigating the quality of life of migrant construction workers in Shenyang, China was also tested using the Split-half reliability method (Qu et al., 2009). 1125 questionnaires obtained from the workers were tested for reliability. To undertake the Split-half reliability method test, the 36 questions/items were split into odd and even ones for all the questionnaires obtained from the workers (Ajayi, 2017). Thereafter, the scores of the responses of odd number questions were correlated with those of even numbered questions to obtain the correlation coefficient r for each split separately and comparing the two, thereby calculating the reliability of each part of the split questionnaire (Qu et al., 2009). It was corrected using the Spearman–Brown prediction formula $[r = 2r_1/(1 + r_1)]$ (Sarmah & Hazarika, 2012), which generated the value of $r = 0.798$ ($P < 0.001$), showing that this questionnaire was relatively stable.

Another study aimed at developing a new individual earthquake resilience questionnaire employed the split-half reliability method (Jiang et al., 2021). The questionnaire to be developed contains 15 questions/items broadly divided into four factors namely, health status, mental resilience, social adaptation and disaster response capacity. Subsequently, the 15 questions/items questionnaire obtained from a total of 952 urban residents affected by Wenchuam earthquake in Dujiangyan City in China were split into odd and even numbered questions. Thereafter, the Spearman correlation of the scores of the responses of odd number questions with those of even numbered questions was calculated to obtain

Table 1.3 Split-half reliability test results

Split-half reliability (n = 952)		
Dimension	Number of items	Split-half coefficient
Health status	3	0.74**
Mental resilience	5	0.88**
Social adaptation	5	0.82**
Disaster response capacity	4	0.83**

**P < 0.01

Source: Jiang et al. (2021)

the correlation coefficient and the split-half reliability of each factor as shown in Table 1.3. The split-half reliabilities of the four factors were 0.85, 0.93, 0.90 and 0.90, respectively, which were considered acceptable.

1.2.2.2 Cronbach alpha

The Cronbach alpha is used to measure the internal consistency of a set of items/variables measuring a construct/concept. Therefore, it measures the degree to which the different items/variables, especially those that each yield numerical response (Lam et al., 2010), but measuring the same construct/concept attains consistent results (Thanasegaran, 2009). The scores on the items/variables designed to measure the same construct/concept should be highly correlated (Thanasegaran, 2009). Therefore, Cronbach's alpha is a function of the average intercorrelations of items and the number of items in the scale (Kimberlin & Winterstein, 2008; Mohajan, 2017). Of note is that having multiple items to measure a construct/concept aids in the determination of the reliability of measurement and, in general, improves the reliability or precision of the measurement (Kimberlin & Winterstein, 2008). Instruments with questions that have more than two responses can be used in this test (Heale & Twycross, 2015), but the greater the number of items in a summated scale, the higher Cronbach's alpha tends to be (Kimberlin & Winterstein, 2008). The Cronbach's alpha result is a number between 0 and 1. An acceptable reliability score is one that is 0.7 and higher (Heale & Twycross, 2015).

Arditi et al. (2013) conducted a study on the managerial competencies of individuals working as managers in the Swedish construction industry and demonstrated the reliability of the questionnaire used through the Cronbach's alpha test of the internal consistency of the questions/items contained. The Cronbach's alpha produces the estimate of reliability coefficient or the appropriateness of item(s) that measure a single unidimensional construct (Davcik, 2014; Ekolu & Quainoo, 2019). The questionnaire comprised of 160 questions/items broadly categorized into 20 key competencies which makes it 8 item per competency. The questionnaire was administered to 112 managers (comprising 44 women and 68 men) who assessed themselves in the competencies by responding to the

160 questions/items based on 5-point Likert scale. The scale comprised of "strongly agree", "agree", "neutral", "disagree" or "strongly disagree" and was converted to a numerical scale where each answer is worth between one and five points. The Cronbach's alpha formula is presented in Equation below; Where N represents the item numbers, σ_i^2 is the variance of the item i and σ_t^2 represents the total variance of the scale (Davcik, 2014).

$$\propto = \left(\frac{N}{N-1}\right) 1 - \left(\frac{\Sigma_{i=1}^{N}\sigma_i^2}{\sigma_t^2}\right) \qquad (1.1)$$

As shown in Table 1.4, 17 of the 20 competencies have a Cronbach's alpha higher than 0.60 and the mean value of the Cronbach's alpha for the 20 competencies is 0.69, which means that the questionnaire that was used for assessing the managerial competencies of managers in Swedish construction industry is reliable (Arditi et al., 2013; Ekolu & Quainoo, 2019).

Another study that identified the key reasons for adopting PPP for construction projects in Ghanaian construction industry also demonstrated the reliability of the questionnaire used through the Cronbach's alpha test of the internal consistency of the questions/items contained (Robert et al., 2014). The uniqueness of this study is the low number of industry practitioners who assessed a

Table 1.4 Cronbach's alpha test results

Key competency	Cronbach's alpha
Initiative	0.74""
Risk taking	0.73""
Innovation	0.80""
Flexibility/Adaptability	0.35
Analytical thinking	0.74""
Decision making	0.74""
Planning	0.77""
Quality focus	0.67""
Oral communication	0.88""
Sensitivity	0.79""
Relationships	0.70""
Teamwork	0.70""
Achievement	0.71""
Customer focus	0.53
Business awareness	0.67""
Learning orientation	0.71""
Authority/Presence	0.83""
Motivating others	0.23
Developing people	0.79""
Resilience	0.74""

"" = > 0.60

Source: Arditi et al. (2013)

17 questions/items questionnaire according to a Likert scale from 1 to 5 (1: not important and 5: extremely important). There were 45 industry practitioners involved. After the Cronbach's alpha test, the estimate of reliability coefficient was 0.838, which signified that the questionnaire was reliable. It is interesting that a low estimate of reliability can result from a small number of questions/items in a questionnaire (Lam et al., 2010).

1.2.2.3 Kuder-Richardson

According to Sarmah and Hazarika (2012), the Kuder-Richardson method was introduced by Kuder-Richardson, a psychometrist, in 1937. The Kuder-Richardson method is like the split-half method except that it is used to measure the degree of internal consistency of items consisting of only two responses (e.g. yes or no, 0 or 1) in a measuring instrument. The method assumes that all items of a test are of equal or almost equal difficulty and intercorrelated (Sarmah & Hazarika, 2012). The common Kuder-Richardson method formula is known to be Kuder-Richardson formula 20 or KR20, which was later simplified to be Kuder-Richardson formula 21 or KR21 (equation shown below). Their difference is that KR21 can produce a direct estimation of reliability using a minimal dataset with only the number of test items, mean and variance as shown in Equation (1.2) (Ekolu & Quainoo, 2019).

$$KR21 = \frac{N}{N-1}\left(1 - \frac{\bar{\chi}(N-\bar{\chi})}{N \cdot \sigma^2}\right) \qquad (1.2)$$

where:
  $\bar{\chi}$ – the mean of all results or scores;
  N – the number of test items or questions;
  σ – variance of all results or scores.

According to Heale et al. (2015), it is calculated by the average of all possible split-half combinations and a correlation between 0 and 1 is generated. Like the split-half method, strong correlations indicate high reliability, while weak correlations indicate the instrument may not be reliable (Kaji & Lewis, 2008). In applying the KR formula, it is assumed that all the test items are of the same level of difficulty. KR21 gives reliability index values lying between 0 and 1, as does Cronbach's alpha (Ekolu & Quainoo, 2019).

Kaji and Lewis (2008) assessed the reliability of 94 items/questions of Johns Hopkins Agency for Healthcare Research and Quality's (AHRQ) drill performance tool by applying it to multiple hospitals in Los Angeles County that participated in the California State-wide disaster drill in November 2005. The study involved 32 fourth-year medical student observers who were deployed to specific zones (incident command, triage, treatment and decontamination) in the participating hospitals. Altogether, the observers completed questions/items specific to their hospital zone using dichotomous rating scale (i.e. better

Table 1.5 Kuder-Richardson test results

Tool component	Number of items	Number of observers	Internal reliability
Incident command zone	45	10	0.97
Triage zone	17	10	0.97
Treatment zone	26	7	0.72
Decontamination zone	18	5	0.92

Total number of items evaluated = 200; $P < 0.05$

preparedness vs poorer preparedness). Based on the scores obtained, the Kuder-Richardson method was used to calculate the degree of internal consistency of the zone-specific evaluative questions in each module of the evaluation tool. The Kuder-Richardson coefficients, by zone, are high. As shown in Table 1.5, they ranged from 0.72 ($P < 0.05$) in the treatment zone to 0.97 ($P < 0.05$) in the incident command zone.

Another study by Heerman et al. (2016) on the parents' perceptions of the safety of their built environments for their preschool-age children in Tennessee, United States employed the Kuder-Richardson method to assess the internal consistency of the questionnaire used. The study involved 610 parents who assessed perceived availability, condition and safety of the built environment with its self-reported use for physical activity through a questionnaire. The instrument consists of 9 items that assess the walkability, bikeability, traffic safety, lighting and crime safety of a participant's neighbourhood. The parents responded to three response options (disagree, don't know/am not sure, agree), and was scored as follows: unsafe = 0, don't know/am not sure = 1 and safe = 2. A composite score was created as a sum of the 9 items, with higher scores representing safer perceived built environments. The resultant scale had a range of 0–18 and demonstrated good internal reliability among items with a Kuder-Richardson coefficient of 0.75.

1.2.3 Equivalence

Equivalence establishes the extent to which the measuring instrument collects information in a consistent manner. According to Heale et al. (2015), equivalence is established by evaluating the consistency among (1) responses of multiple users of an instrument (inter-rater reliability) and (2) among alternate forms of an instrument (parallel-form or alternate-form reliability). Often, observational instruments or rating scales are developed to evaluate the behaviours of subjects who are being directly observed. However, any measure that relies on the judgments of raters or reviewers requires evidence that any independent, trained expert would come to the same conclusion (Kimberlin & Winterstein, 2008). It is useful because human observers will not necessarily interpret answers the same way; raters may disagree as to how well certain responses or material demonstrate knowledge of the construct being assessed (Mohajan, 2017).

1.2.3.1 Inter-rater reliability

The more that individual judgment is involved in a rating, the more crucial it is that independent observers agree when applying the scoring criteria (Kimberlin & Winterstein, 2008). Inter-rater reliability establishes the equivalence of ratings obtained with a measuring instrument when used by different raters (Mohajan, 2017). Therefore, it is used to determine the level of agreement between two or more raters (Heale & Twycross, 2015; Kimberlin & Winterstein, 2008). On the other hand, intra-rater reliability establishes the equivalence of ratings obtained with a measuring instrument used by a single rater over a period (Fink, 2010; Su et al., 2014). An example of intra-rater reliability can be found in Weich et al. (2001). An example of inter-rater reliability in research is when respondents are asked to give a score for the relevancy of each item on an instrument. The consistency in their scores relates to the level of inter-rater reliability of the instrument (Heale & Twycross, 2015). The reliability is determined by the correlation of the scores from two or more independent raters, or the coefficient of agreement of the judgments of the raters (Mohajan, 2017). For categorical variables, Cohen's kappa is commonly used to determine the coefficient of agreement when two raters or observers classify events or observations into categories based on rating criteria (Kimberlin & Winterstein, 2008). Where there are more than two ordered categories, the weighted kappa statistic can also reflect the degree of difference between raters.

(Weich et al., 2001)

Weich et al. (2001) conducted a study to develop and perform inter-rater reliability of a 27 questions/items Built Environment Site Survey Checklist (BESSC). Of the 27 questions/items in the checklist, 25 of them have fixed categorical responses, while the remaining two required the researcher to rank features of the built environment (like the proportion of space used in particular ways), and to estimate the distance from the centre of the "housing area" to a range of amenities (like the bus stop). The research design involved the allotment of each housing area an identifying number (and there were 11 of them), and every third area in the experimental ward was selected for reliability study. The 11 areas were rated independently by two postgraduate student researchers in urban design and architecture using the BESSC. Also, one of the students used the BESSC to rate the other 75 housing areas across the two wards. In line with Kaji and Lewis (2008) and Su et al. (2014), one of the authors in Weich et al. (2001) assessed the checklist together with the student researchers to ensure consistency in their understanding of the questions/items. After data collection, the inter-rater reliability of the BESSC was assessed using the kappa and Weighted kappa statistics for the categorical variables and item rankings. The degree of inter-rater reliability for the 25 categorical questions/items in BESSC, across the 11 housing areas is presented in Table 1.6. The complete analysis and results can be found in

Table 1.6 Weighted kappa statistics test results

Survey items	Number of categories	Kappa
Housing form	2	0.54
Number of storeys of buildings	5	0.57
Type of access to dwellings	3	0.72
Number of dwellings per entrance	8	0.64
Number of dwellings in the housing area	6	0.67
Age of housing	5	0.64
Number of trees in public domain	5	0.65
Nature of space outside dwellings	3	0.61
Private gardens: proportion	3	0.65
Private balconies: proportion	3	0.52
Shared recreational space	2	1.00
Number of pedestrial entrances	5	0.65
Entrances visible from roads	5	0.50
Disused buildings	2	0.62
Evidence of graffiti	3	0.76

Parallel-form reliability (or alternate-form reliability)

Weich et al. (2001), but most of the items in the checklist were of satisfactory inter-rater reliability with kappa coefficients not less than 0.60. The authors interpreted the results to demonstrate substantial reliability.

Parallel-form reliability (or alternate-form reliability) is like test-retest reliability but with an exception that a different (or an alternate) form of the original test is administered at different times (Drost, 2011). According to Heale et al. (2015), the concepts being tested are the same in both versions, but the expressions may be presented differently. As the name implies, two or more versions of the test are constructed that are equivalent in content and level of difficulty, e.g. professors use this technique to create makeup or replacement exams because students may already know the questions from the earlier exam (Thanasegaran, 2009). The measuring instrument used is stable when there is a high correlation between the scores (values) obtained each time the tests are completed (Heale & Twycross, 2015). A low correlation indicates the presence of measurement error, which is construed as using two different scales in both tests (Drost, 2011). For example, when testing for general spelling, one of the two independently composed tests might not test general spelling but a more subject-specific type of spelling such as business vocabulary.

References

Bannigan, K., & Watson, R. (2009). Reliability and validity in a nutshell. *Journal of Clinical Nursing*, 18(23), 3237–3243.

Brink, H. I. (1993). Validity and reliability in qualitative research. *Curationis*, 16(2), 35–38.

Dainty, A. (2008). Methodological pluralism in construction management research. *Advanced Research Methods in the Built Environment*, 1, 1–13.

Thanasegaran, G. (2009). Reliability and validity issues in research. *Integration & Dissemination*, 4, 35–40.

Ajayi, B. K. (2017). A Comparative Analysis of Reliability Methods. *Journal of Education & Practice*, 8(25), 160–163.

Arditi, D., Gluch, P., & Holmdahl, M. (2013). Managerial competencies of female and male managers in the Swedish construction industry. *Construction Management and Economics*, 31(9), 979–990.

Davcik, N. S. (2014). The use and misuse of structural equation modeling (SEM) in management research: A review and critique. *Journal of Advances in Management Research*, 11(1), 47–81.

Drost, E. A. (2011). Validity and reliability in social science research. *Education Research and perspectives*, 38(1), 105–123.

Ekolu, S. O., & Quainoo, H. (2019). Reliability of assessments in engineering education using Cronbach's alpha, KR and split-half methods. *Global Journal of Engineering Education*, 21(1), 24–29.

Fink, A. (2010). Survey research methods. University of California, Los Angeles, Los Angeles, CA, USA; The Langley Research Institute, Pacific Palisades, CA, USA. Retrieved on June 1, 2021 from EXA20.pdf

Heale, R., & Twycross, A. (2015). Validity and reliability in quantitative studies. *Evidence-based Nursing*, 18(3), 66–67.

Heerman, W. J., Mitchell, S. J., Thompson, J., Martin, N. C., Sommer, E. C., Van Bakergem, M., ... & Barkin, S. L. (2016). Parental perception of built environment characteristics and built environment use among Latino families: a cross-sectional study. *BMC Public Health*, 16(1), 1–8.

Jiang, N., Cheng, J., Ni, Z., Ye, Y., Hu, R., & Jiang, X. (2021). Developing a new individual earthquake resilience questionnaire: A reliability and validity test. *PLoS One*, 16(1), e0245662.

Kaji, A. H., & Lewis, R. J. (2008). Assessment of the reliability of the Johns Hopkins/Agency for Healthcare Research and Quality hospital disaster drill evaluation tool. *Annals of Emergency Medicine*, 52(3), 204–210.

Kimberlin, C. L., & Winterstein, A. G. (2008). Validity and reliability of measurement instruments used in research. *American Journal of Health-System Pharmacy*, 65(23), 2276–2284.

Lam, P. T., Chan, E. H., Poon, C. S., Chau, C. K., & Chun, K. P. (2010). Factors affecting the implementation of green specifications in construction. *Journal of Environmental Management*, 91(3), 654–661.

Mohajan, H. K. (2017). Two criteria for good measurements in research: Validity and reliability. *Annals of Spiru Haret University. Economic Series*, 17(4), 59–82.

Robert, O. K., Dansoh, A., & Ofori–Kuragu, J. K. (2014). Reasons for adopting public–private partnership (PPP) for construction projects in Ghana. *International Journal of Construction Management*, 14(4), 227–238.

Sarmah, H. K., & Hazarika, B. B. (2012). Determination of Reliability and Validity measures of a questionnaire. *Indian Journal of Education and Information Management*, 5(11), 508–517.

Su, M., Du, Y. K., Liu, Q. M., Ren, Y. J., Kawachi, I., Lv, J., & Li, L. M. (2014). Objective assessment of urban built environment related to physical activity—development, reliability and validity of the China Urban Built Environment Scan Tool (CUBEST). *BMC Public Health, 14*(1), 1–10.

Qu, B., Guo, H. Q., Liu, J., Zhang, Y., & Sun, G. (2009). Reliability and validity testing of the SF-36 questionnaire for the evaluation of the quality of life of Chinese urban construction workers. *Journal of International Medical Research, 37*(4), 1184–1190.

Weich, S., Burton, E., Blanchard, M., Prince, M., Sproston, K., & Erens, B. (2001). Measuring the built environment: validity of a site survey instrument for use in urban settings. *Health & Place, 7*(4), 283–292.

2 Understanding validity in research

Ayokunle Olanipekun, Vian Ahmed,
Alex Opoku and Monty Sutrisna

2.1 Validity test

Validity is defined as the extent to which an instrument measures what it purports to measure (Kimberlin & Winterstein, 2008) and does so cleanly without accidentally including other factors (Thanasegaran, 2009). For instance, does cost performance measure project performance? Does prequalification predict contractor competence? Therefore, when a measuring instrument is used to obtain information and consequently tested, the emphasis of validity is not necessarily on the items or the scores of the items, but the interpretations that can be extrapolated from the test scores relative to the construct/concept being measured (Kimberlin & Winterstein, 2008). To be valid, the inferences made from scores need to be meaningful (Drost, 2011; Mohajan, 2017). For example, a survey (a measure or measuring instrument) designed to explore depression among construction workers, which measures their anxiety levels, would not be considered valid, and vice versa (Heale and Twycross, 2015). This example demonstrates the inextricable link between validity and reliability. Despite that it is designed to measure depression, if the instrument measures and produces anxiety scores repeatedly, then it is reliable. Therefore, a measuring instrument that reliably measures something other than what is intended is not valid (Kimberlin & Winterstein, 2008). A valid instrument must be reliable, while a reliable instrument may not necessarily be valid (Thanasegaran, 2009). The violations of the validity of an instrument critically impact the function and functioning of a testing instrument at a much greater level than the violations of reliability of such measuring instrument (Kimberlin & Winterstein, 2008; Thanasegaran, 2009). According to Taherdoost (2016), there are four categories of validity: face validity, content validity, construct validity and criterion validity, as presented in Table 2.1.

2.1.1 Face validity

Face validity is a quick assessment of what a test is measuring (Drost, 2011; Mohajan, 2017). For items/variables of measure in questionnaire, face validity is the degree to which the items/variables relate to the underlying construct/concept

DOI: 10.1201/9780429243226-3

Table 2.1 Categories of validity

Categories of validity	Description
Face validity	The extent to which items/variables relate to the underlying construct/concept from the face value
Content validity	The extent to which a research instrument accurately measures all aspects of a construct
Construct validity	The extent to which a research instrument (or tool) measures the intended construct
Criterion validity	The extent to which a research instrument is related to other instruments that measure the same variables

Source: (Heale and Twycross, 2015; Taherdoost, 2016)

(Taherdoost, 2016). Face validity is determined based on the judgement of non-experts such as test takers (Taherdoost, 2016), which means a test has face validity if its content simply looks relevant to the person taking the test (Heale and Twycross, 2015). For this reason, face validity is considered subjective and the least scientific type of validity (Drost, 2011; Mohajan, 2017). For instance, one might look at a measure of reading ability, read through the paragraphs, and decide that it seems like a good measure of reading ability (Drost, 2011). By looking at a measure (or measuring instrument on the surface), a researcher might be able to tell whether all the needed questions are asked and whether the appropriate language and language level has been used (Fink, 2010). Another suggestion for examining face validity by Taherdoost (2016) is to use dichotomous scale that comprises categorical option of "Yes" and "No" to indicate the representativeness of items/variables of constructs/concepts. Subsequently, data collected can be analysed using Cohen's Kappa Index (CKI) in determining the face validity of the instrument (Taherdoost, 2016).

2.1.2 Content validity

Content validity is often equated with the broad concept of "validity" due to definitional similarity, but it is only correct that content validity is one of the categories of validity. Content validity is the extent to which a measure comprehensively assesses the characteristics it is intended to measure (Fink, 2010). For example, a BE researcher interested in developing a measure of "preconstruction performance" has to first define the concept (i.e. "what is preconstruction performance?" "How is it different from postconstruction performance?") and then pen variables/items that adequately contain all aspects of the definition. According to Mohajan (2017), whether a measure comprehensively covers a content area is determined based on the judgement of experts in the field. In practice, selected experts are presented a measure (or measuring instrument) to judge by scoring the content validity (Taherdoost, 2016). The scoring is subjected to statistical analysis to produce content validity ration (CVR) using the following formula.

$$CVR = \frac{n_e - \frac{N}{2}}{\frac{N}{2}}$$

According to Taherdoost (2016), CVR is the content validity ratio, n_e is the number of experts that indicate content validity and N is the total number of experts.

Jiang et al.'s (2021) study that was previously mentioned (in chapter 1, section 1.2.2) also analysed the content validity of the new individual earthquake resilience questionnaire that was developed. The questionnaire contains 15 questions/items broadly divided into four factors, namely, health status, mental resilience, social adaptation and disaster response capacity. After administering the questionnaire and obtaining the scores from 952 urban dwellers who were supposedly experts on the subject in China, the content validity of the questions/items relative to the concept of earthquake resilience was evaluated. The item content validity ratio (I-CVR) ranged from 0.87 to 1, and the scale average content validity ratio (S-CVR/Ave) was 0.94, indicating good content validity.

Another study by Brennan & Cotgrave (2013) aimed to develop a measure to quantify attitudes towards social, economic and environmental dimensions of sustainability in construction tested for the content validity of the measure. Following the identification of 70 potential measures of the dimensions in the literature (de Barros Ahrens et al., 2020), 10 experts from both academia and industry were selected to assess the contents in terms of relevance to the concepts/dimensions. Specifically, the experts were asked to choose and score questions/items that they thought were worded in a way that would elicit strong attitudinal responses on the scale of 1–5 (1 being weak and 5 being strong). Eventually, 7 of the 10 experts rated the questions/items as instructed. In what seems like a departure, the inter-item correlation of the expert ratings was carried out. The premise of decision of content validity was that questions/items should inter-correlate at a significant level (either 5% or 1%) if they measure aspects of the same thing. The authors reported that all the 70 individual questions/items correlate significantly as do the dimensions as shown in Table 2.2.

2.1.3 Criterion (or concrete validity) validity

The criterion validity of a (test) measure (or measuring instrument) is the degree of correspondence between the test measure and one or more external referents (criteria), usually measured by their correlation (Drost, 2011; Thanasegaran, 2009). This type of validity provides evidence about how well scores on a new or an initial measure correlate with other measures of the same construct or very similar underlying constructs that theoretically should be related (Kimberlin & Winterstein, 2008). This means that an actual/initial measure (or measuring instrument) is tested such as surveying employees in a company to report their

Table 2.2 Inter-correlation test results

Correlated items	Environmental subscale	Social subscale	Economic subscale
Environmental subscale	1.00		
Social subscale	0.54	0.51	
Economic subscale	0.37		0.53

Correlation is significant @ <0.05 and <0.01

Source: Brennan & Cotgrave (2013)

salaries (Drost, 2011). To obtain the criterion validity of the survey, the actual salary records of the employees can be obtained and correlated with their reported salaries (Drost, 2011). In both cases, the concept being measured is the salary and the salary record is the criterion used to correlate the survey report of the employees to determine the validity of measure (or measuring instrument) (Heale and Twycross, 2015; Taherdoost, 2016). To correlate both measures (e.g. salary reports and salary records) is to compare them to determine the degree of correlation. Therefore, the criterion (measure) must be standard and be established as a valid measure to provide a strong basis of comparison. According to Bannigan and Watson (2009), criterion validity involves comparing the scale being developed with a criterion measure that has been established as valid (Bannigan and Watson, 2009). Criterion validity is measured in two ways: (1) concurrent validity and (2) predictive validity (Heale and Twycross, 2015).

2.1.3.1 Concurrent validity

Concurrent validity (of a measure) is tested when the criterion is obtained at the same time (or concurrently) as the test scores (Bannigan and Watson, 2009). It means there is no extended period between the testing measure and the criterion measure. For instance, a new simple test is to be used in place of an old troublesome one, which is considered useful; measurements are obtained on both at the same time (Mohajan, 2017). The aim is often to examine a measure's ability to reflect current or present ability (Drost, 2011; Thanasegaran, 2009). For instance, in a situation in which a new instrument has some advantage over the gold standard measure, such as an increased ease of use or reduced time or expense of administration. These advantages would justify the time and effort involved in the development and validation of a new instrument. An example of such a situation is a researcher developing a self-administered version of an instrument that had been validated for person-to-person interviewer administration (Kimberlin & Winterstein, 2008). In establishing concurrent validity, scores on an instrument are correlated with scores on another (criterion) measure of the same construct or a highly related construct that is measured concurrently in the same subjects.

(Kimberlin & Winterstein, 2008)

Larsson et al. (2019) aimed to investigate the criterion validity of the SED-GIH – a single item/question tool with categorical answering options for assessing sitting time sedentary behaviour. The authors observed that while sedentary behaviour (especially sitting time) has been linked to metabolic and cardiovascular health, and, disease prevention, the present tools for assessing it more likely overestimate sitting than the single item/question tool which often underestimate sitting time. At the same time, according to the authors, categorical answering options (such as SED-GIH) have been recommended. Also, the activPAL is a wearable technology for measuring sedentary behaviours such as sitting time. Therefore, the SED-GIH was administered to 284 middle-aged adults in Stockholm and Gothenburg, Sweden, who also wore the activPAL to obtain data about their sitting time sedentary behaviour in 2016–2017. Sitting time is the criterion measure in the study. Following the data obtainment, the Spearman's rho was used to test the correlation between the SED-GIH and activPAL-SIT data with 95% confidence interval (CI). The results showed that the SED-GIH question correlated moderately with activPAL-SIT (rho = 0.31) and was interpreted as low criterion validity.

2.1.3.2 Predictive validity

The predictive validity is like concurrent validity except that there is a time elapse between the criterion and test measures. It means the criterion occurs at a future time or the criterion measurement is obtained at some time after the administration of the test (Drost, 2011). In other words, the scores from the predictor measure are taken first and then the criterion data is collected later (Taherdoost, 2016). Therefore, what is evaluated is the ability of the test to accurately predict the criterion (Kimberlin & Winterstein, 2008). For instance, surrogate outcomes such as blood pressure and cholesterol levels are based on their predictive validity in projecting the risk of cardiovascular disease (Mohajan, 2017). Furthermore, it is also the ability of an assessment tool to predict future performance on another assessment of the same construct (Taherdoost, 2016). This means that it assesses the operationalization's ability to predict something it should theoretically be able to predict (Taherdoost, 2016). For example, a score of high self-competence in project planning should predict the likelihood a participant completing a project plan (Heale and Twycross, 2015). The degree of correlation between the criterion variable and scores on the testing instrument is to assert good criterion validity (Drost, 2011). Validity coefficients are ultimately derived from the correlation between these components. From this, one can calculate a coefficient of determination for the measures by squaring the validity coefficient (Thanasegaran, 2009). The higher the correlation between the criterion and the predictor indicates, the greater the predictive validity. If the correlation is perfect, that is, 1, the prediction is also perfect. Most of the correlations are only modest, somewhere between 0.3 and 0.6 (Mohajan, 2017). Spittaels et al. (2010) reported the predictive validity of a questionnaire that was developed for assessing the physical activity-related environmental factors among Europeans. The questionnaire contains both 49-items/questions and 11-items/questions on physical activity and fitness, respectively. To undertake the predictive validity testing process, the questionnaire was completed by 190 adults. One week after, the physical activity of the adults was assessed using an accelerometery and the data obtained. Thereafter, the predictive validity was examined by correlating the results of the questionnaires with the physical activity data from the accelerometery. The Pearson correlations between the scores of the environmental variables and accelerometer data show significant correlation and ranged from 0.19 to 0.38. It was reported to indicate moderate validity.

2.1.4 Construct validity

Construct validity is how well a concept, an idea or a behaviour can be transformed into a functioning or operating reality (Drost, 2011; Taherdoost, 2016). Practically, constructs are not directly observable. To observe them, items/ variables are developed as representations/operationalizations of constructs (Thanasegaran, 2009). Therefore, the operationalisation of a construct often involves a series of items/variables that are hypothesized to correspond to the latent construct (Long & Johnson, 2000; Thanasegaran, 2009). The construct validity of a measure is therefore concerned with the theoretical relationship of a variable (e.g. a score on some scale) to other variables (Thanasegaran, 2009). Therefore, it answers the question whether inferences about test scores related to a concept under investigation can be drawn (Heale and Twycross, 2015). For instance, if a construction professional has a high score on a survey that measures of digital competence, does this professional truly have a high degree of digital competence? The evaluation of construct validity can be best understood through two construct-validation processes: first, testing for convergence across different measures (or convergent validity) and second, testing for divergence between measures (or divergent validity) (Taherdoost, 2016).

2.1.4.1 Convergent validity

Convergent validity refers to the degree to which two measures of constructs that theoretically should be related, are in fact related (Taherdoost, 2016). Therefore, convergent validity should arise when considering two constructs hypothesized to be related (Thanasegaran, 2009). A strong convergent validity is established when the scores obtained with two different instruments measuring the same concept are highly correlated (Heale and Twycross, 2015; Mohajan, 2017). Correlational evidence is performed by testing a priori hypotheses developed about how the measurement under development will correlate with another measurement scale (Bannigan and Watson, 2009). The testing of hypotheses formulated about the measurement scales the measure will correlate with is known as 'convergent validity' (Bannigan and Watson, 2009). Also, factor analysis can be used to test convergent validity of measures. Items that load above 0.40, which is the minimum recommended value in research are considered for further analysis. Also, items cross loading above 0.40 should be deleted.

(Taherdoost, 2016)

de Barros Ahrens et al. (2020) evaluated the convergent validity of a 50 questions/items instrument for assessing working environment as a single tool based on quality of life (QL), quality of work life (QWL) and organizational climate (OC). The Exploratory Factor Analysis (EFA) and Person Moment correlation methods were used for the evaluation of the instrument. Factor analysis is a statistical method used to test the structural validity of a scale and describes variability among observed variables in terms of fewer unobserved variables called factors (Qu et al., 2009). The research design involved the surveying of 229 workers in a Beta company. Following the obtainment of data, the Kaiser–Meyer–Olkin (KMO) index was used to verify the suitability of the application of EFA for the data obtained, and it returned a KMO index of 0.917. According to the authors, it demonstrated a strong suitability. Also, the Bartlett's sphericity test was significant ($\chi^2 = 7465.349$, Df = 1225, $p \leq 0.000$), which further indicated the suitability of the EFA. Subsequently, the Principal Component Analysis (PCA) involving a varimax rotation was used to estimate the factor loadings and specificity. The number of factors estimated was determined by the Scree plot through retention of components with eigenvalues greater than 1 (Qu et al., 2009). The analysis revealed that ten factors with eigenvalues greater than 1 should be retained. As shown in Table 2.3, the retained factors were named as pointed by the instrument's theoretical construct.

To evaluate the convergent validity of the instrument using Pearson Moment correlation, the correlation matrix demonstrated the existing relationships between the pairs of questions/items with positive correlation results. The authors reported that this is an indication of the convergent validity of the instrument. Of the 50 questions/items, eight of them presented low correlation ($r = 0.26$–0.49),

Table 2.3 Principal factors retained after analysis

Principal components/factors
1 – Health
2 – Emotional and psychological
3 – Spiritual
4 – Sleep and rest
5 – Work and life
6 – Work conditions
7 – Leadership management
8 – Remuneration and functional assistance
9 – Functional responsibility
10 – Personal relations and company image

Source: de Barros Ahrens et al. (2020)

thirty-eight presented moderate correlation ($r = 0.7-0.89$) and four presented high correlations ($0.9-1.0$) (Davcik, 2014).

Jiang et al.'s (2021) study that was previously mentioned (in chapter 1, section 1.2.2) also analysed the convergent validity of the new individual earthquake resilience questionnaire that was developed. The questionnaire contains 15 questions/items broadly divided into four factors, namely, health status, mental resilience, social adaptation and disaster response capacity. The research design involved the administration of the questionnaire to 952 urban dwellers who were supposedly experts on the subject in China. Following the obtainment of data, the EFA was carried out. According to Qu et al. (2009), the EFA is a multivariate analysis method for examining the convergence of a scale by describing the variability among observed variables in terms of fewer unobserved variables known as factors. The cumulative variance contribution rate is an indication of the extent that the common factors explain the total variance, and the factor loadings indicate the correlations between a variable/question/item and a common factor (Jiang et al., 2021). To assess the suitability of the EFA for the data obtained, KMO returned an index of 0.95 which indicated a strong suitability (Qu et al., 2009). Furthermore, using the PCA method, the 15 questions/items/variables that were divided into four factors had a cumulative contribution rate of 74.97%. As shown in Table 2.4, all the questions/items/variables loaded significantly onto the four components/factors with loading factors higher than 0.5. Therefore, they converged acceptably into four factors/components.

Table 2.4 Factor loadings after rotation

Items	Component/ Factor 1	Component/ Factor 2	Component/ Factor 3	Component/ Factor 4
1				0.64
2				0.85
3				0.79
4		0.71		
5		0.77		
6		0.73		
7		0.74		
8		0.63		
9			0.64	
10			0.71	
11			0.71	
12			0.74	
13			0.56	
14	0.81			
15	0.81			
16	0.81			
17	0.82			

Source: Jiang et al. (2021)

2.1.4.2 Discriminant validity

Discriminant validity is obtained when comparing a measure (scale of interest) to an unrelated construct (a construct that is not hypothesized to be related with the scale of interest) and there is when there is little to no correlation (Thanasegaran, 2009). Often, it follows theoretical postulation that two constructs are uncorrelated, and the scores obtained by measuring them are indeed empirically found to be so, that is, to differentiate one group from another (Mohajan, 2017). The implication for measurement is that a measuring instrument might is poorly correlated to instruments that measure different variables. For example, there should be a low correlation between an instrument that measures motivation and one that measures self-efficacy of construction site workers (Heale and Twycross, 2015). Therefore, the constructs that should have no relationship do, in fact, not have any relationship (Taherdoost, 2016).

Olanipekun (2018) investigated 150 project owners to determine their motivation for delivering green building projects in the Australian construction industry. Undertaken the investigation, the authors reviewed the literature to identify a list of seven motivation factors that were neither scientifically structured nor confirmed (See Table 2.5). Following the obtainment of data from the project owners, the EFA comprising of the principal axis factoring (PAF) and oblique rotation was used to reveal the latent structure of the data and uncovering the common factors in the process. Subsequently, the confirmatory factor analysis (CFA) and model fitting were carried out to validate the factors that were derived from the EFA. As reported in Olanipekun et al. (2018), two factors that comprise three and four items/variables respectively were confirmed (See Table 2.5).

Meanwhile, following the confirmation of the factors, it was necessary to evaluate the extent to the respective questions/items/variables of one factor does not measure another or the discriminant validity (Davcik, 2014). To assess the discriminant validity, the average variance extracted (AVE) for the factor was calculated. Discriminant validity is affirmed when AVE of a factor is greater than its squared correlations with other factors (Fornell and Larcker, 1981). The results of the discriminant validity test showed that the structure has

Table 2.5 Confirmed variables after factor analysis

Items/variables	Factors
Improved quality of life (M1)	**Internal motivation (INT)**
Pro-environmental altruism (M2)	
Enhanced reputation (M3)	
Persuasive influence (M4)	
Market appeal (M5)	**External motivation (EXT)**
Financial incentives (M6)	
Non-financial incentives (M7)	

Source: Olanipekun et al. (2018)

dissimilar constructs for the two factors because the AVE of factor 1 (0.649) and factor 2 (0.736), are greater than the squared factor correlation between INT and EXT ($R^2 = 0.230$) (Hon et al., 2012).

2.2 Internal validity

Internal validity refers to the extent to which research findings are a true reflection or representation of reality rather than being the effects of extraneous variables (Brink, 1993). For example, a manager in a construction company tests the employees on leadership satisfaction, and only 50% of them responded to the survey and all of them liked their boss. Does this mean that the manager has a representative sample of employees or a bias sample? (Drost, 2011). Given that there is a relationship, is the relationship a causal one? Are there no confounding factors that influence the finding? Therefore, internal validity speaks to the validity of the research itself. There are many threats to internal validity of a research, and they include: history, maturation, testing, instrumentation, selection, mortality, diffusion of treatment and compensatory equalization, rivalry and demoralization (Drost, 2011). To enhance internal validity, the researcher can describe appropriate strategies, such as triangulation, prolonged contact, member checks, saturation, reflexivity and peer review (Mohajan, 2017).

2.3 External validity

External validity of a study (or relationship) implies generalizing to other persons, settings and times (Drost, 2011). External validity addresses the degree or extent to which such representations or reflections of reality are legitimately applicable across groups (Brink, 1993). Of note is that generalizing to well-explained target populations should be clearly differentiated from generalizing across populations. The former is critical in determining whether any research objectives which specified populations have been met, and the latter is crucial in determining which different populations have been affected by a test to assess how far one can generalize (Cook & Campbell, 1979) cited in (Drost, 2011). For example, can the causal relationship observed in a construction site be replicated in a public institution, in a bureaucracy, or on a military base? (Drost, 2011). According to Mohajan (2017), external validity can be increased: i) achieving representation of the population through strategies, such as, random selection, ii) using heterogeneous groups, iii) using non-reactive measures and iv) using precise description to allow for study replication or replicate study across different populations, settings, etc.

References

Bannigan, K., & Watson, R. (2009). Reliability and validity in a nutshell. *Journal of Clinical Nursing*, 18(23), 3237–3243.

Bannigan, K., Boniface, G., Nicol, M., Porter-Armstrong, A., Scudds, R., & Doherty, P. (2009). The nature and value of research priority setting in healthcare: case study of the POTTER project. *Journal of Management & Marketing in Healthcare*, 2(3), 293–304.

Brink, H. I. (1993). Validity and reliability in qualitative research. *Curationis*, *16*(2), 35–38.

Brennan, M., & Cotgrave, A. J. (2013). Development of a measure to assess attitudes towards sustainable development in the built environment: a pilot study. In Proceedings of 29th Annual ARCOM Conference (pp. 2–4).

Davcik, N. (2014). The use and misuse of structural equation modeling (SEM) in management research: A review and critique. *Journal of Advances in Management Research*, *11*(1), 47–81.

de Barros Ahrens, R., da Silva Lirani, L., & de Francisco, A. C. (2020). Construct validity and reliability of the work environment assessment instrument WE-10. *International Journal of Environmental Research and Public Health*, *17*(20), 7364.

Cook, T. D. and Campbell, D. T. (1979), *Quasi-Experimentation: Design and Analysis Issues for Field Settings*, Chicago: Rand McNally.

Drost, E. A. (2011). Validity and reliability in social science research. *Education Research and Perspectives*, *38*(1), 105–123.

Fornell, C., & Larcker, D. F. (1981). Evaluating structural equation models with unobservable variables and measurement error. *Journal of Marketing Research*, *18*(1), 39–50.

Heale, R., & Twycross, A. (2015). Validity and reliability in quantitative studies. *Evidence-Based Nursing*, *18*(3), 66–67.

Jiang, N., Cheng, J., Ni, Z., Ye, Y., Hu, R., & Jiang, X. (2021). Developing a new individual earthquake resilience questionnaire: A reliability and validity test. *PLOS One*, *16*(1), e0245662.

Kimberlin, C. L., & Winterstein, A. G. (2008). Validity and reliability of measurement instruments used in research. *American Journal of Health-System Pharmacy*, *65*(23), 2276–2284.

Larsson, K., Kallings, L. V., Ekblom, Ö., Blom, V., Andersson, E., & Ekblom, M. M. (2019). Criterion validity and test-retest reliability of SED-GIH, a single item question for assessment of daily sitting time. *BMC Public Health*, *19*(1), 1–8.

Long, T., & Johnson, M. (2000). Rigour, reliability and validity in qualitative research. *Clinical Effectiveness in Nursing*, *4*(1), 30–37.

Mohajan, H. K. (2017). Two criteria for good measurements in research: Validity and reliability. *Annals of Spiru Haret University. Economic Series*, *17*(4), 59–82.

Qu, B., Guo, H. Q., Liu, J., Zhang, Y., & Sun, G. (2009). Reliability and validity testing of the SF-36 questionnaire for the evaluation of the quality of life of Chinese urban construction workers. *Journal of International Medical Research*, *37*(4), 1184–1190.

Spittaels, H., Verloigne, M., Gidlow, C., Gloanec, J., Titze, S., Foster, C., ... & De Bourdeaudhuij, I. (2010). Measuring physical activity-related environmental factors: reliability and predictive validity of the European environmental questionnaire ALPHA. *International Journal of Behavioral Nutrition and Physical Activity*, *7*(1), 1–19.

Taherdoost, H. (2016). Validity and reliability of the research instrument; how to test the validation of a questionnaire/survey in a research. How to test the validation of a questionnaire/survey in a research. *International Journal of Academic Research in Management* *5*(3): 28–36.

Thanasegaran, G. (2009). Reliability and validity issues in research. *Integration & Dissemination*, *4*, 35–40.

Part II
Reliability test: Research case study examples

3 An investigation into contributing factors of excess inventory within the cosmetic industry in the UAE

An AHP analysis as form of inter-rater reliability

Vian Ahmed, Sara Saboor, Heba Khlaif and Dana Yazbak

3.1 Background

An organization's operational strategy greatly depends on its Supply Chain Management (SCM) which acts as a significant part of the organization and enables it to thrive by retaining its market share. Therefore, to reduce costs organizations must constantly reassess their supply chain and adopt sustainable practices such as Green Supply Chain Management strategies and gain a competitive advantage in the market. However, one of the major drawbacks to the supply chain management or green supply chain management is the failure to address inventory management practices, whereby inadequate planning and ineffective inventory management of supply chain contributes to carrying costs of inventory and harmful environmental impacts.

According to Nnamdi (2018), "Excess Inventory" is an indication of demand-supply mismatch often described in terms of operational liability to the organization storage space, handling cost and working capital, where the existence of inventory can be justified, it effectively serves as a working stock, safety stock, anticipation stock, pipeline stock or a decoupling stock. However, any item inventory that doesn't fulfil this purpose can be classified as "Excess Inventory", which is classified by three categories: Dead stock, degraded stock and slow-moving stock (Toelle and Tersine, 1989). Accordingly, organizations face enormous challenges on how to manage their excess and dead inventory due to the lifecycle of their stock. Therefore, the long lifecycle of these equipment require organizations to ensure the provision and availability of the replacements until equipment termination. At the same time these organizations need to control their inventory carrying cost and inventory write-offs in order to ensure that only needed inventory/stock are available for their consumers (Nnamdi, 2018). Despite, the efforts as identified by Vereecke and Verstraeten (1994), the proportion of the stock range that is devoted to excess inventory is often significant with the irregular demand occurrence connected with such items and the consequential low contribution to the total turnover of an organization, these slower moving SKUs can establish up to 60% of the total stock value.

DOI: 10.1201/9780429243226-5

Thus, several challenges have been pointed out, given the complication and significance of inventory management. As such, Boone et al. (2008) argue that companies often lack a system assessment, suffer the weakness of supply chain relationships and the inaccuracy of demand forecasts. In addition, there are number of factors contribute towards excess inventory that not only have a negative effect on the organization's financial health such as blocking the capital, limitation of space, scarce resources, high maintenance cost and also cause harmful impact to the environment. Although the factors contributing to excess inventory are widely spread, no rating schemes exist in the literature that differentiate the most occurring factors to the least occurring factors and can be adopted by the industry to set a benchmark of where to begin when tackling excess inventory, hence, the issue of excess inventory and problems associated acts as the biggest threat to any industry. This study is mainly focussed on the cosmetic industry which according to the report conducted by Zero Waste Europe alone contribute to about 142 billion units of plastic waste in 2018 and high carbon footprint from transferring ingredients to distribution and excess inventory. The industry that worth over $500 billion where countries like India ranked eight with total sales of $14 billion, China $62 billion and Middle East and Africa (MEA) region's worth $32 billion in 2018, where the statistics of the industry estimated to grow to $820 billion by year 2023, is also concerning in terms of increase in excess inventory and its related environmental effects (Mint, 2021).

This study identifies the major factors that contribute to excess inventory within the cosmetic industry in the UAE, by adopting AHP analysis which due to its multiple independent assessors results in superior form of inter-rater agreement level, i.e. inter-rater reliability.

3.2 Excess inventory in the UAE cosmetic industry

According to McManus (2010), UAE is known to be a hub to many different multi-cultural businesses around the world and has adopted a successful reputation for trade. Furthermore, he also states that the "UAE traders are amongst the most optimistic in the world. This is partly due to the improved and more stable oil prices, which enables higher government spending, creating a positive chain reaction across the regional economies". McManus continues to explain that UAE has become the third largest re-export hub after Singapore and Hong Kong which explains the desire for many companies wanting to carry out business in the region.

Amongst the largest industries in the UAE, is the cosmetic industry which provides a wide range of products including skin, hair and oral care, cosmetics and other niche segments was the leading product segment in the marketplace and holding 43.5% share of the market in 2018 (Kumar, 2005). This industry has increased its focus on natural skincare behaviours, non-invasive beauty processes and the rising influence of beauty and lifestyle bloggers which has added to a steady growth in the UAE's beauty sector. According to a report by Euromonitor

International in 2018, it is estimated that the retail value of the Middle East and Africa (MEA) region's beauty and personal care market will be worth $35.9 billion in 2018 and will continue to grow at a compound annual growth rate of nearly 10% over the next four years. Customers in the UAE spent $247 per capita on cosmetics and individual care, more than any other country in the Middle East, and ninth worldwide; this is forecast to grow to $294 in 2020 (Sadaqat, 2020). According to Mordor Interlliehnce (2020), the UAE cosmetics market is projected to grow at a Compound Annual Growth Rate (CAGR) of over 4% during 2019–2024. Due to the increasing demand from youth population, growth in working women population, increasing adoption of western culture and lifestyle and increasing number of beauty salons are aiding UAE cosmetics market which further highlights the need for more research concerning the cosmetic industry within the UAE.

Given that this sort of industry is very subjective, stockpiling is expected. Thus, it becomes crucial to optimize inventory control to help reduce inventory waste (excess inventory). However, before implementing any model and technique it is important to determine the major contributing factors to access inventory. As such, this study underpinned from literature, the factors of that contribute to excess inventory as shown in Table 3.1, to further explored within the UAE cosmetic industry.

Table 3.1 Literature review findings

Factors	Authors
Demand variation/Changes in market condition	(Crandall and Crandall, 2003; Dureno, 1995; Dasu et al., 2012; McCullen and Towill, 2002)
Supply variation/Bulk purchase	(Crandall and Crandall, 2003; Dureno, 1995; Jaggi et al., 2011)
Internal variation/Inadequate planning and execution systems	(Toelle and Tersine, 1989; Crandall and Crandall, 2003; Luo et al., 2020)
Sales and marketing	(Crandall and Crandall, 2003; Lummus et al, 2003)
Engineering/Changes in design and specification	(Crandall and Crandall, 2003; Dureno, 1995, Romanov, 2013)
Production/Master schedule smoothing	(Toelle and Tersine, 1989; Crandall and Crandall, 2003; Shin et al., 2019)
Accounting/Finance	(Crandall and Crandall, 2003; Lutilsky, 2018; Magsi, 2012)
Forecasting errors	(Toelle and Tersine, 1989, Dureno, 1995; Cao and Shen, 2019)
Inventory record Inaccuracies/Poor stock management	(Toelle and Tersine, 1989; Dureno, 1995, Eminue et al., 2019)
Long or variable lead times	(Toelle and Tersine, 1989, Dominguez et al., 2019)
Obsolescence	(Toelle and Tersine, 1989; Nnamdi, 2018)
Distribution channel adjustments/ Item relocation	Toelle and Tersine, 1989; Dureno, 1995, Barratt et al., 2018)
Changes in inventory holding costs	(Toelle and Tersine, 1989)
Returns from customers	(Dureno, 1995)

3.3 Research methodology

The focus of this section is on the methodological approach adopted to identify the highest contributing factors to "Excess Inventory" and their significant impact on the cosmetic industry in the UAE.

> *Step I: Literature review* – The literature reviewed earlier in this study aided with the identification of 14 major causes of excess inventory.
>
> *Step II: Semi-structure interviews* – A set of semi-structured interviews were held, targeting a panel of experts from the cosmetic industry in the UAE was targeted in order to validate and tailor the findings from literature review in relation the UAE cosmetic industry. The semi-structured interviews were conducted online to help validate the factors that contribute to excess inventory in the UAE cosmetic industry. As such the questions were setup to identify the profile of the participants, the participants' perception of the impact of Excess Inventory in the Supply Chain Cycle, and to validate and identify any missing factors in relation to the excess inventory in the cosmetic industry.
>
> *Step III: Focus group* – This step adopts a focus group approach consisting of experts from the cosmetic industry with the intent to rank the most important factor based on their significant impact on the excess inventory in the cosmetic industry in the UAE.

The next section discusses the results and analysis of the adopted approach.

3.4 Results and analysis

This section highlights the results of the semi-structure interviews and focus group which aided in identifying the major causes of excess inventory in the UAE cosmetic industry.

3.4.1 Semi-structured interviews

A sample of four experts were therefore targeted, with 6 to 18 years of experience and different key positions in the cosmetic industry such as a Business Planning manager, Demand & Supply planning manager, Systems, Applications and Products (SAP) Key user and an Operation Director as summarized in Table 3.2.

The semi-structured interviews aided in understanding the participants' perception of the impact of Excess Inventory in the Supply Chain Cycle, and to

Table 3.2 Participant's profile

Participants	Title	Experience
PI	Business planning manager	8 years
PII	Demand & supply planning manager	6 years
PIII	SAP key user	6 years
PIV	Operation director	18 years

validate and identify any missing factors in relation to the excess inventory in the cosmetic industry, where to evaluate the interviewees' perception of the impact of excess inventory within the cosmetic industry they were asked to; **identify the harmful effects of retaining Excess Inventory in the Supply Chain Cycle within the cosmetic industry in the UAE.**

In response to this question, the participants generally agreed that excess inventory in the UAE cosmetic industry are high and mainly contribute to additional organizational costs due to fluctuations in demands. In addition, it was found that the excess inventory has cost implication and harmful impact on the environment.

Furthermore, to validate the excess inventory factors in the cosmetic industry a set of 14 factors identified from the literature (summarized in column 2 of Table 3.3a) were presented to the participants while **asked to give their expert opinion to whether they believed any of these factors contributed to the Excess Inventory within the cosmetic industry in the UAE.** All the participants were in agreement that "Demand Variation, Forecast Accuracy, Sales (Realistic) and Marketing (Dreamers), Engineering/Changes in Design & Specifications, Production/Master Schedule Smoothing and Returns from Customers are relevant while, Supply Variation/Bulk Purchase, Accounting/Finance, Inventory Record Inaccuracies/Poor Stock Management, Long or Variable Lead Times, Obsolescence, Distribution Channel Adjustments/Item Relocation, and Changes in Inventory Holding Costs" are not relevant to the UAE cosmetic industry, as shown in Table 3.3a. While factor of Forecasting error highly correlates with Forecast Accuracy, thus they have been combined together.

The participants were also asked to whether they **believed that there are any other contributing Excess Inventory factors that occur within UAE's cosmetic industry Supply Chain Cycle.** Participant IV suggested **Stock Cover** to be a

Table 3.3a Participants perception summary

Main causes of excess inventory		Participants views
1	Demand variation/Change in market conditions	Relevant to cosmetic industry
2	Supply variation/Bulk purchase	Not relevant to cosmetic industry
3	Internal variation/Inadequate planning and execution systems – Forecast accuracy – Launches	Relevant to cosmetic industry
4	Sales (realistic) and marketing (dreamers)	Relevant to cosmetic industry
5	Engineering/Changes in design and specifications	Relevant to cosmetic industry
6	Production/Master schedule smoothing	Relevant to cosmetic industry
7	Accounting/Finance	Not relevant to cosmetic industry
8	Forecasting errors (Point 3)	Adding to factor 3
9	Inventory record inaccuracies/Poor stock management	Not relevant to cosmetic industry
10	Long or variable lead times	Not relevant to cosmetic industry
11	Obsolescence	Not relevant to cosmetic industry
12	Distribution channel adjustments/Item relocation	Not relevant to cosmetic industry
13	Changes in inventory holding costs	Not relevant to cosmetic industry
14	Returns from customers	Relevant to cosmetic industry

Table 3.3b Summary of the findings from interviews

Main causes of excess inventory
1   Demand variation/Change in market conditions
2   Internal variation/Inadequate planning & execution systems – Forecast accuracy – Launches
3   Sales (realistic) and marketing (dreamers)
4   Engineering/Changes in design & specifications
5   Production/Master schedule smoothing
6   Returns from customers
7   Stock cover
8   Global crisis

contributing factor and stating "If a company considers five months of SKUs to be healthy if it has an SKU at 4.5 months so it is considered acceptable. If the stock cover policy changes to make the healthy SKU at three months so there is an excess of 1.5 months of excess inventory". In addition, the participants confirmed that "recession, pandemics, natural disasters are a major impact that affects market conditions leading to affecting marketing demand". Therefore, **Global Crisis** can be considered as a contributing factor.

Table 3.3b shows the summary of the findings which present the experts' opinion of the relevant factors that impact on the Excess Inventory of the UAE cosmetic industry. As such, amongst the 14 factors identified from the literature, six factors will be considered by this study as the most relevant factors that impact on Excess Inventory in the UAE cosmetic industry. In addition two more factors were added as a result of the interviews as shown in Table 3.3b.

These factors will be further adopted evaluated by a group of experts in order to rank them based on their significant impact on the excess inventory in the cosmetic industry in the UAE.

3.4.2 Focus group

The intent of this step is to rank the most important factors based on their significant impact on the excess inventory in the cosmetic industry in the UAE. A focus group was therefore formed by targeting six experts with 1 to 4 years of experience working as Demand planner, Customer care, Business Analyst and Sales coordinator. Table 3.4 presents a summary of the participants' profile.

The participants were then briefed of on the Excess Inventory factors identified through interviews and were asked to suggest any missing factors **that are likely to contribute to Excess Inventory within the cosmetic industry in the UAE. The** participants agreed that **"Catalogue Rationalization"** can be an addition factor whereby, the more products in an organization's catalogue, the more difficult it is to get an accurate forecast on all the Stock Keeping Unit (SKU)s, which could also make the forecast accuracy is low which in return can generate more excess inventory. Thus, a total of nine factors will be considered for the next step

Table 3.4 Focus group participants profile

Sample	Title	Experience
Participant 1	E-commerce customer care executive	1 Year
Participant 2	Demand planner	4 Years
Participant 3	Demand planner	5 Years
Participant 4	Business analyst	1 Year
Participant 5	Customer care returns executive	3 Years
Participant 6	Sales and operations coordinator	3 Years

of this study using the "Analytical Hierarchy Process" in order to measure the importance of one factor in comparison to the others (Figure 3.1). This will aid with the decision-making process in complex environment where there exists a number of variables and criteria that need prioritization in order to come up with alternative solutions for reducing excess inventory.

3.4.3 Analytical hierarchy process (AHP) – inter-rater reliability

The Analytical Hierarchy Process (AHP) is a form of analysis conducted to analyze complex decisions containing three essential parts such as the ultimate goal or problem, possible solutions such as alternatives and criteria that the alternative will be judge on. The analysis allows multiple stakeholders (i.e. the rater) to provide with their subjective judgement on the pairwise comparison of criteria. However, despite the widespread popularity of rating scales and methods, the literature reports on lack of focus of researchers on the reliability of their rating data whose quality is critical to the success of their research. Therefore, it is pertinent to establish the reliability of the set of ratings to validate that the variance in the ratings is the result of a systematic rather than a random ordering of the objects (Howard, 2000).

Thus, the benefit of adopting AHP analysis is its multiple independent assessors that results in superior form of inter-rater agreement level, i.e. inter-rater reliability which provides a measure of the degree of agreement among independent individuals (i.e. raters). As adopted by Ben-Assuli et al. (2020) that compares the

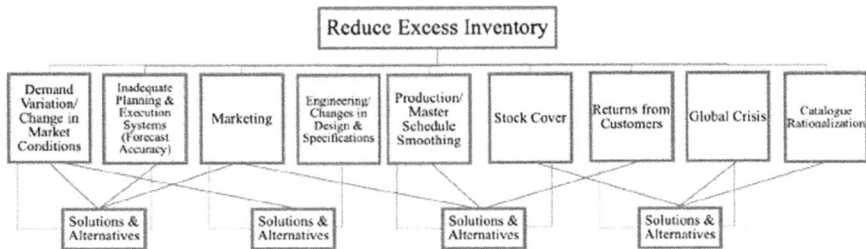

Figure 3.1 Analytical hierarchy process

inter-rate agreement level for the AHP analysis to the Likert type scale and concluded that AHP-based scales yield higher agreement levels in term of internal consistency and Cronbach alpha (Reliability).

The participants of the focus group were asked to rank factors based on their pairwise comparison which is followed by aggregating their individual judgement in the AHP matrix as adopted by Kouatli (2019) that defines the participants of the focus group as a new "individual" as they behaves like one. Thus, Ramanathan and Ganesh (1994) suggest that the mutuality requirement for the judgements must be satisfied, therefore, for this reason the geometric mean is adopted to aggregate their individual judgements rather than an arithmetic mean. The AHP analysis was conducted using the following steps:

A. Step 1 - AHP matrix

Let suppose n alternatives $(A_i, i = 1, ..., n)$ and r number of decision makers $(D_k, k = 1, ..., r)$, let $A^{[k]} = (a_{ij}^{[k]})$ be the judgement matrix provided by the k-th decision maker when comparing n elements $(i, j = 1, ..., n)$ and let β_k be the weight that the k-th decision maker $(k = 1, ..., r)$ has in forming the group decision. In this case, the weight of k-th decision maker is considered as equal (Vereecke and Verstraeten, 1994; Figure 3.2).

$$a_{ij}^{[G]} = \prod_{k=1}^{r} 1 \left(a_{ij}^{[k]} \right)^{\beta}{}_k \tag{3.1}$$

Note that: $a_{ij} = 1/a_{ji}$, for $i \neq j$, and $a_{ii} = 1$, all i.

B. Step 2 - Normalization

This step involves the normalization of the pairwise comparison matrix by using the following equations (Vereecke and Verstraeten, 1994) as shown in Figure 3.3.

$$a_{ij} = a_{ij} / \sum_{i=1}^{m} a_{ij}, m \text{ - dimensional column vector} \tag{3.2}$$

Geometric Mean Matrix	1	2	3	4	5	6	7	8	9
1	1	0.5	5	6	4	1	1	0.5	1
2	2	1	7	8	9	4	1	2	2
3	0.2	0.142857143	1	1	4	0.25	0.5	0.166666667	1
4	0.166666667	0.125	1	1	5	0.2	1	0.2	0.166666667
5	0.25	0.111111111	0.25	0.2	1	0.142857143	0.166666667	0.5	0.166666667
6	1	0.25	4	5	7	1	1	1	0.5
7	1	1	2	1	6	1	1	0.5	0.25
8	2	0.5	6	5	2	1	2	1	2
9	1	0.5	1	6	6	2	4	0.5	1

Figure 3.2 Analytical hierarchy process (AHP) analysis

Normalized Matrix	1	2	3	4	5	6	7	8	9
1	0.033697047	0.006624606	0.130039012	0.14507772	0.06122449	0.012804097	0.030837004	0.058823529	0.013866878
2	0.26957638	0.052996845	0.182054616	0.14507772	0.163265306	0.115236876	0.008810573	0.058823529	0.083201268
3	0.006739409	0.007570978	0.026007802	0.103626943	0.081632653	0.012804097	0.008810573	0.058823529	0.013866878
4	0.004813864	0.007570978	0.00520156	0.020725389	0.081632653	0.007682458	0.008810573	0.058823529	0.011885895
5	0.011232349	0.006624606	0.006501951	0.005181347	0.020408163	0.006402049	0.008810573	0.058823529	0.011885895
6	0.101091142	0.017665615	0.078023407	0.103626943	0.12244898	0.038412292	0.008810573	0.058823529	0.016640254
7	0.067394095	0.370977918	0.182054616	0.14507772	0.142857143	0.268886044	0.061674009	0.058823529	0.016640254
8	0.303273427	0.476971609	0.234070221	0.186528497	0.183673469	0.345710627	0.555066079	0.529411765	0.74881141
9	0.202182285	0.052996845	0.156046814	0.14507772	0.142857143	0.19206146	0.308370044	0.058823529	0.083201268

Figure 3.3 Normalization

C. Step 3 - Calculate eigenvectors and max eigenvalues

The step helps to calculate eigenvectors and the max eigenvalue which is considered as Lemda as shown in Figure 3.4.

> **Definition 1:** An eigenvector of a square matrix A is a vector v such that $A \times v = \lambda v$
>
> **Definition 2:** An eigenvalue is the scalar λ associated with an eigenvector v

D. Step 4 - Consistency test

The consistency property of the AHP matrix was checked to ensure that there exists consistency of judgement in pairwise comparison. The Consistency index and Consistency ratio can be calculated as shown in (Vereecke and Verstraeten, 1994) Table 3.5:

$$CR = \frac{CI}{RI} \quad \text{and} \quad CI = \frac{\lambda_{max} - n}{n-1} \tag{3.3}$$

where N is the number of elements and criteria that have being compared, whereas RI is the random index. The consistency ratio of AHP analysis is the measure of the consistency of the judgements in the given evaluation matrix relative to the large sample of random judgements.

Factor Number	Factor	Eigenvector (Weight)
1	Change in Market Conditions	5.478%
2	Forecast Accuracy	11.989%
3	Marketing	3.554%
4	Engineering	2.302%
5	Production	1.510%
6	Stock Cover	6.062%
7	Returns from Customers	14.604%
8	Global Crisis	39.595%
9	Catalogue Rationalization	14.907%

Figure 3.4 Weigths for excess inventory factors

Table 3.5 Consistency test

Consistency determinants	Calculated
Lambda max	10.3412914
Consistency index (CI)	0.16766142
Consistency ratio (CR)	0.11562857

The consistency ratio less than or equal to 0.1 is considered as acceptable judgement as defined by Saaty. However, for this study the consistency ratio was found to be 0.115, which is still perceived to acceptable as Park and Kim (2014), who performed their own study of consistency verification and claimed that if Consistency Ratio (CR) is less than 0.1 then this reflects complete consistency, if CR is less than or equal to 0.2 then the consistency is permissible, but they should definitely be re-investigated if it is greater than 0.2. Another source by Pauer et al. (2016) performed a study to measure the influence of individual or group judgements and the impact of using geometric and arithmetic mean on their AHP analysis. One of the study's findings is that a CR ≤0.2 proved to be an acceptable level of consistency. Thus, the following sources ensured that the final consistency ratio achieved is of an acceptable level as it was ≤0.2. This implies that the judgement and results of Analytical Hierarchy process (AHP) in this study was found to be acceptable.

Furthermore, the results presented in Figure 3.5 shows the most significant factors that cause excess inventory in the cosmetic industry in the UAE, whereby **Forecast accuracy** acts as a major contributor towards the excess inventory followed by **Global Crisis** (e.g. COVID 19).

Therefore, it can be concluded from the consistency ratio that there exists an acceptable degree of agreement among independent individuals (i.e. raters), which indicates higher agreement and inter rater reliability.

Factors	Focus Group		Amended	
	Weights	Ranking	Weights	Ranking
1: Change in Market Conditions	5.48%	6	11.94%	4
2: Forecast Accuracy	11.99%	4	24.46%	1
3: Marketing	3.55%	7	4.80%	7
4: Engineering	2.30%	8	4.30%	8
5: Production	1.51%	9	2.45%	9
6: Stock Cover	6.06%	5	11.47%	5
7: Returns from Customers	14.60%	3	9.86%	6
8: Global Crisis	39.59%	1	16.00%	2
9: Catalogue Rationalization	14.91%	2	14.72%	3

Figure 3.5 AHP results

3.5 Main findings and conclusions

Inventory management is considered as a critical component with a significant impact on the organization financial performance, as it involves determining the right balance of production and storage of products. Such balance determines the right quantity of products to order and hold which otherwise either results in shortage of supply and interruption of sales, poor consumer relation or an increase of tied up capital, deterioration, damage, loss and obsolescence to the inventory increasing the *excess Inventory*. Unfortunately, the cosmetic industries are no exception to this despite being one of the largest and supply chains in the UAE. This study therefore adopts the inter-rater reliability as a mean of AHP to ensure the quality of the rating is withheld, with the variance in the ratings is systematic and the measure of the consistency of the judgements for the analysis is considered acceptable. Following an in-depth analysis of the qualitative and quantitative results obtained, this paper identified *Forecast Accuracy, Global Crisis and Catalogue Rationalization* as the most important factors that contribute to excess inventory in this industry.

References

Barratt, M., Kull, T.J. and Sodero, A.C., 2018. Inventory record inaccuracy dynamics and the role of employees within multi-channel distribution center inventory systems. *Journal of Operations Management*, 63, pp. 6–24.

Ben-Assuli, O., Kumar, N., Arazy, O. and Shabtai, I., 2020. The use of analytic hierarchy process for measuring the complexity of medical diagnosis. *Health Informatics Journal*, 26(1), pp. 218–232.

Boone, C.A., Craighead, C.W. and Hanna, J.B., 2008. Critical challenges of inventory management in service parts supply: A Delphi study. *Operations Management Research*, 1(1), pp.31–39.

Cao, Y. and Shen, Z.J.M., 2019. Quantile forecasting and data-driven inventory management under nonstationary demand. *Operations Research Letters*, 47(6), pp. 465–472.

Crandall, R.E. and Crandall, W.R., 2003. Managing excess inventories: A life-cycle approach. *Academy of Management Perspectives*, 17(3), pp. 99–113.

Dasu, S., Ahmadi, R. and Carr, S.M., 2012. Gray markets, a product of demand uncertainty and excess inventory. *Production and Operations Management*, 21(6), pp. 1102–1113.

Dominguez, R., Cannella, S., Ponte, B. and Framinan, J.M., 2020. On the dynamics of closed-loop supply chains under remanufacturing lead time variability. *Omega*, 97, p. 102106.

Dureno, D.J., 1995. Inventory management–a business issue. *Hospital Materiel Management Quarterly*, 17(2), pp. 6–11.

Eminue, U.O., Titus, C.U. and Udo, L.O., 2019. Stock management strategies and safeguarding of inventory shrinkage in large-scale retail outlets in akwa ibom state, Nigeria: An empirical review. *European Journal of Economic and Financial Research*. DOI:10.46827/EJEFR.VOI0.575

Jaggi, C.K., Khanna, A. and Verma, P., 2011. Two-warehouse partial backlogging inventory model for deteriorating items with linear trend in demand under inflationary conditions. *International Journal of Systems Science*, 42(7), pp. 1185–1196.

Kouatli, I., 2019. People-process-performance benchmarking technique in cloud computing environment: An AHP approach. *International Journal of Productivity and Performance Management.* 69(9), ISSN: 1741-0401

Kumar, S., 2005. Exploratory analysis of global cosmetic industry: Major players, technology and market trends. *Technovation, 25*(11), pp. 1263–1272.

Lummus, R.R., Duclos, L.K. and Vokurka, R.J., 2003. The impact of marketing initiatives on the supply chain. *Supply Chain Management: An International Journal.* Vol. 8 No. 4, pp. 317–323. ISSN: 1359-8546

Luo, L., Jin, X., Shen, G.Q., Wang, Y., Liang, X., Li, X. and Li, C.Z., 2020. Supply chain management for prefabricated building projects in Hong Kong. *Journal of Management in Engineering, 36*(2), p. 05020001.

Lutilsky, I.D., Liović, D. and Marković, M., 2018. Throughput accounting: Profit-focused cost accounting method. *Interdisciplinary Management Research Xiv (Imr 2018), 14,* pp. 1382–1395.

Magsi. H., 2012. Green supply chain management. *Pakistan & Gulf Economist, 31*(24), pp. 67–69.

McManus. S. 2010. MENA trade & supply chain: Holding fast in the Middle East. *Trade Finance,* pp 797–800.

McCullen, P. and Towill, D. 2002). Diagnosis and reduction of bullwhip in supply chains. *Supply Chain Management, 7*(3), pp. 164–179.

Mint, 2021. "The Ugly Side of beauty waste". Available at: https://www.livemint.com/mint-lounge/features/unseen-2019-the-ugly-side-of-beauty-waste-11577446070730.html

Mordor Intelligence., 2020. "UNITED ARAB EMIRATES COSMETIC PRODUCTS MARKET - GROWTH, TRENDS, COVID-19 IMPACT, AND FORECASTS (2021–2026)". Available at https://www.mordorintelligence.com/industry-reports/united-arab-emirates-cosmetics-products-market-industry

Nnamdi, O., 2018. Strategies for managing excess and dead inventories: A case study of spare parts inventories in the elevator equipment industry. *Operations and Supply Chain Management: An International Journal, 11*(3), pp. 128–138.

Park, B. and Kim, R.Y., 2014. Making a decision about importance analysis and prioritization of use cases through comparison the analytic hierarchy process (AHP) with use case point (UCP) technique. *International Journal of Software Engineering and Its Applications, 8*(3), pp. 89–96.

Pauer, F., Schmidt, K., Babac, A., Damm, K., Frank, M. and von der Schulenburg, J.M.G., 2016. Comparison of different approaches applied in analytic hierarchy process–An example of information needs of patients with rare diseases. *BMC Medical Informatics and Decision Making, 16*(1), pp. 1–11.

Ramanathan, R. and Ganesh, L.S., 1994. Group preference aggregation methods employed in AHP: An evaluation and an intrinsic process for deriving members' weightages. *European Journal of Operational Research, 79*(2), pp. 249–265.

Romanov, A., 2013. Analysis and reduction of excess inventory at a heavy equipment manufacturing facility (Doctoral dissertation, Massachusetts Institute of Technology).

Sadaqat. R. 2020. UAE consumers among top spenders in cosmetics, personal care globally. Khaleej Times. https://www.khaleejtimes.com/business/retail/uaeconsumers-among-top-spenders-in-cosmetics-personal-care-globally.

Shin, M., Lee, H., Ryu, K., Cho, Y. and Son, Y.J., 2019. A two-phased perishable inventory model for production planning in a food industry. *Computers & Industrial Engineering, 133,* pp. 175–185.

Tinsley, H. E. A., & Weiss, D. J. (2000). Interrater reliability and agreement. In H. E. A. Tinsley & S. D. Brown (Eds.), *Handbook of applied multivariate statistics and mathematical modeling* (pp. 95–124). Academic Press. DOI:10.1016/B978-012691360-6/50005-7

Toelle, R.A. and Tersine, R.J., 1989. Excess inventory: financial asset or operational liability. *Production and Inventory Management Journal*, 30(4), pp. 32–35.

Vereecke, A. and Verstraeten, P., 1994. An inventory management model for an inventory consisting of lumpy items, slow movers and fast movers. *International Journal of Production Economics*, 35(1–3), pp. 379–389.

4 An investigation into underpinning criteria of "subjective happiness" in an academic environment – a parallel form of reliability

Sara Saboor, Alia Al Sadawi, Malick Ndiaye and Vian Ahmed

4.1 Background

Happiness as the term stands have been perceived is a positive emotion and the satisfaction of an individual with life. Several authors argued that happiness does not imply an abundance of health, beauty, money and good fortune. As such, happiness does not depend on external factors but a condition prepared and cultivated by an individual and their satisfaction with life whether good or bad (Csikszentmihalyi, 2013).

One of the most popular and accepted definitions of happiness is by Alexandrova (2005), who defined happiness as a sum of an individual's current emotional state that comprises of the feeling of joy, excitement, anger, love, hope, sadness and amusement which refers to the experience of an individual's present and past that contribute to the overall satisfaction. Though several definitions existed over the years by a number of authors, however, the broad concept of happiness can be classified into two categories such as objective happiness and subjective happiness.

According to Frey and Stutzer (2006), *Objective Happiness* represents the decision utility (Instant Utility) of an individual who can be measured through scans of brain activities, whereas, *Subjective Happiness* represents the "experienced utility of an individual's life that can be measured through the self-administrative test based on the surveys and questionnaires, that test the satisfaction of the individual with the past experiences and variables of its life. Measures of subjective happiness are often to determine the social capital of societies that enables them to develop and flourish.

Therefore, this study focuses on understanding the concept of subjective happiness and its criteria, the importance of which is evident from the literature and is reported by a number of studies that recommend the adoption of this construct in the policymaking process and a number of fields such as Philosophy, Economics, Psychiatry and Neurosciences (Extremera & Fernández, 2014). For example, Easterlin (1974) and Scitovsky (1976) were one of the earliest studies linking happiness with economics, sparking interest amongst economists to measure and categorize the determinants of subjective happiness and wellbeing. With education being an important contributor to the growth and development

DOI: 10.1201/9780429243226-6

of society, it becomes essential to understand and adopt the measure of subjective happiness to enhance its benefits.

In addition, literature reports on a number of studies such as (Tabbodi et al., 2015), (Kiamarsi, and Momeni (2013), and Applasamy et al (2014), which mostly focused on the impact of students' happiness on their motivation and academic achievement. For example, Applasamy et al (2014) suggests that students with high happiness factors can score high and perform well. However, little or no studies have been conducted to identify the underpinning criteria and importance of these criteria on Subjective Happiness within academic environments. It is, therefore, within the intentions of this study to understand "Subjective Happiness", it scales and identifies the underpinning criteria of Subjective Happiness of students in an academic environment.

4.2 Literature review or subjective happiness

Subjective Happiness is often translated as subjective wellbeing where the individual self-evaluates their satisfaction with the quality of life and wellbeing obtained through surveys and questionnaires. Subjective happiness is also an essential variable for the welfare economy and a guide for policymaker, which ultimately results in a flourishing community (Zulkifli, 2013). This section, therefore, presents the most validated and reliable instruments reported in the literature to measure happiness.

A significant number of instruments are reported by literature to measure happiness such as Affect Balance Scale (Bradburn, 1969), Affectometer (Kammann & Flett, 1983), Affective Intensity Measure (Larsen, 1984), Global Happiness Scale (Fordyce, 1977), Positive and Negative Affect Schedule (PANAS) (Watson, Clark, & Tellegen, 1988). The main idea behind these instruments is that happiness is the frequency associated with the effect: High-positive affect (PA) and low negative affect (NA). However, the idea was rejected through implementing empirical evidence by authors such as (Larsen & Prizmic, 2008) who suggested that people who face adverse events can show a high level of happiness depending on the situation and process of adaptation. In addition, Lyubomirsky and Boehm (2010) defined the concept of hedonic adaptation and ruled out the direct relationship between happiness and positive affect.

However, in 1999, the first non-theoretical-based approach was introduced by Lyubomirsky and Lepper (1999). The research presented a "Subjective" measure of happiness. The scale was titled a Subjective Happiness Scale (SHS); it was the first scale that measures happiness without considering what happiness is. Similarly, (Lyubomirsky & Lepper, 1997) used this approach to measure happiness in college students. The data was collected from a total of 2,732 participants from the United States (two college campuses and one high school campus), the Adult community in two of the California cities and Russia. The research was based on the SHS. Five measures of happiness and wellbeing were used to validate SHS, i.e. Affect Balance Scale; The Delighted-Terrible Scale; The Global Happiness Item; The Recent Happiness Item and the Satisfaction With Life Scale.

These validated and reliable instruments were adopted by various studies in different areas to determine "Subjective Happiness". Though the scales have been adopted in the field of academic environment the focus of these studies was to evaluate the subjective happiness of the students and staff on their academic achievement. But few or no studies focus on identifying the underpinning criteria of Subjective Happiness in an academic environment. This study, therefore, intends to identify the underpinning criteria of Subjective Happiness in an academic environment and determine the students' perception of the importance of these criteria towards their happiness.

4.2.1 Underpinning criteria of subjective happiness

A number of studies focus on various criteria that impact student satisfaction with the academic environment which in turn aids them to improve their academic achievement.

One of the studies by (Henry, 2004) focused his research on 315 university students in Regina to measure their life satisfaction or "Subjective Happiness". The measures proposed by the study were: 5-item Satisfaction with Life Scale (SWLS) and Satisfaction with specific aspects of life. The criteria for the satisfaction among university students used were "School Performance", "Courses Taking", "School Facilities", "Instructors' Quality of Teaching", "Relationships with Close Friends", "Relationship with Father", "Relationship with Mother", "Relationships with Siblings", "Relationship with Spouse/Partner/Significant Other", "Physical Appearance", "Self-image", "Leisure or Recreational Activities", "Financial Security", "Material Possessions/Comfort", "Physical Health", "Living Environment", "Living Arrangements", "Job Situation", "Social Life", "School Life" and "Spiritual Life".

Likewise, Chan et al. (2005) assessed the satisfaction students using multiple instruments to determine the happiness/satisfaction in university students such as the Satisfaction with Life Scale (SWLS) developed by Diener et al. (1985), Satisfaction with specific aspects of life by Chow (2005) and The Perceived Stress Scale (PSS; Cohen et al., 1983). The findings of the research indicate that relationships, self-image, physical appearance, income, extracurricular activities, satisfaction with school work (HWM), satisfaction with resources and school environment, relationships formed, time management, health and university reputations were the most important influences on the levels of satisfaction of students.

Moreover, other criteria of Subjective Happiness in an academic environment identified through literature (Crawford and Henry, 2004), (Chan et al., 2005), (Zhang et al, 2012), (Zulkifli, 2013), (Ramzi, 2007), (Applasamy et al (2014), (King and Datu 2012) and (Diana, 2014) were summarized in Table 4.1.

This study intends to adopt the 14 criteria listed in Table 4.1 to determine the student's perception of the importance and significance of these criteria towards their "Subjective Happiness", and their relative importance by adopting the Relative Importance Index (RII) analysis and the Analytical Hierarchy

Table 4.1 Findings of the literature

Criteria	Authors
Campus life	(Crawford and Henry, 2004), (Ramzi, 2007)
Academic quality/ Education	(Crawford and Henry, 2004), (Chan et al., 2005), (Zhang et al, 2012), (Zulkifli, 2013), (Ramzi, 2007), Applasamy et al (2014), (Diana, 2014)
School facilities/ Services	(Crawford and Henry, 2004), (Chan et al., 2005), (Zhang et al, 2012), (Ramzi, 2007)
Social life	(Crawford and Henry, 2004), (Chan et al., 2005), (Zhang et al, 2012), King and Datu 2012 7, (Medvedev and Landhuis 2018)
Relationship (family/ significant other)	(Crawford and Henry, 2004), (Zulkifli, 2013), (Diana, 2014), King et al., 2014)
Relationship with peers	(Crawford and Henry, 2004) (Chan et al., 2005), (Zulkifli, 2013), King et al., 2014)
Self-image/ Achievement	(Crawford and Henry, 2004), (Diana, 2014), King and Datu 2017
Leisure/ Extracurricular activities	(Crawford and Henry, 2004), (Chan et al., 2005), (Zhang et al, 2012), (Zulkifli, 2013), (Diana, 2014)
Security	(Crawford and Henry, 2004)
Health	(Crawford and Henry, 2004), (Chan et al., 2005), (Zhang et al, 2012), (Zulkifli, 2013), Applasamy et al (2014), (Diana, 2014), King and Datu 2017, (Medvedev and Landhuis 2018)
Living environment	(Crawford and Henry, 2004), (Ramzi, 2007), Applasamy et al (2014), (Diana, 2014), (Medvedev and Landhuis 2018)
Economic (income and job)	(Crawford and Henry, 2004), (Chan et al., 2005), King and Datu 20127, (Krsmanovic, 2017)
Time management	(Crawford and Henry, 2004), (Chan et al., 2005), (Zulkifli, 2013), Applasamy et al (2014)
Development opp	(Ramzi, 2007), (Crawford and Henry, 2004)
Political	Applasamy et al (2014)

Process (AHP) and measuring the consistency of the findings of both methods as Parallel-form reliability (and alternate-form reliability).

4.3 Methodological approach

Having identified the criteria that underpin "Subjective Happiness" from literature, those being; Campus Life, Academic quality/Education, School facilities/Services, Social Life, Relationship (Family/Significant other), Relationship with peers, Self-Image/Achievement, Leisure/Extracurricular activities, Security, Health, Living Environment, Economic (Income and Job), Time management, Development opportunity and Political affiliation, a set of unstructured interviews were conducted with the experts in the fields to validate the findings from the literature. This study also utilizes the use of RII analysis and AHP by conducting two separate surveys that help determine the student's perception of the importance of the identified criteria and their subjective happiness in an academic environment, using a Higher Education Institution as a case study. In addition, a parallel or alternate form of reliability was adopted to ensure consistency between the findings of both methods.

4.4 Data collection and analysis

4.4.1 Semi-structured interview

To validate the "Subjective Happiness" factors identified from the literature, a set of semi-structured interviews were conducted with the experts in a Higher Education Institute. The experts recommended the addition of two more criteria that they deem suitable as contributing factors to the Subjective Happiness of students. The criteria suggested were: "Sense of Belonging" and "Value for Diversity" as shown in Table 4.2.

To test the reliability of the validated factors in Table 4.2, this study adopts a **Parallel form (Alternate Form) Reliability** to determine the students' perception of the importance of the criteria of Subjective Happiness on their happiness in an academic environment. To enable this, a survey was administered targeting students at the American University of Sharjah (AUS) in the UAE. AUS is an independent non-profit educational institute, known to be in the Top 3 universities in the world for diversity with the highest percentage of international students with different cultures and perceptions of happiness. The institute represents 90 plus nationalities with 51% male and 49% female students. Being a "melting pot of cultural and perception diversity, the institute was an idol choice to determine the student's perception of the importance of criteria identified from literature on their 'Subjective Happiness'".

Table 4.2 Validate underpinning criteria

Criteria
Campus life
Academic quality/Education
School facilities/Services
Social life
Relationship (Family/Significant other)
Relationship with peers
Self-image/Achievement
Leisure/Extracurricular activities
Security
Health
Living environment
Economic (income and job)
Time management
Development opp
Political
Sense of belonging
value for diversity

4.4.2 Questionnaire surveys

This study administered two separate surveys to determine the students' perception of the importance of the underpinning criteria identified in Table 4.2. This section describes the approaches and results obtained to achieve this objective.

4.4.2.1 Relative importance index survey

The RII survey is comprised of three sections; The demographic profile section, followed by the Satisfaction with Life Scale (SWLS) section that allows the respondents to self-evaluate their happiness and satisfaction. The survey also allows the respondents to rank the criteria on a 5 Level Likert Scale where (1 – Not Important to 5 – Highly Important) based on the importance of the defined criteria (Table 4.2) to their subjective happiness.

Relative importance index analysis allows the respondents to rank the criteria based on their relative importance, it captures the perception of respondents in terms of the most important factor as shown in equation:

$$RII = \sum W/(N*A) \qquad (4.1)$$

where,
   W = weighting as assigned on Likert's scale by each respondent in a range from 1 to 5, A = Highest weight (here it is 5), and N = Total number in the sample.

Table 4.3 shows the results of the RII analysis based on the importance of the identified criteria to the students' subjective happiness. For this study criteria that weigh below, 0.50 were removed from the analysis, and the top nine criteria are selected further to conduct AHP analysis.

It can be therefore concluded that the students perceive academic quality, Campus Life, Health, Relationships with Peers, Security, Senses of Belonging, Social Life, Living Environment and Campus Facilities are criteria that of relative importance to their Happiness; whereas, Self-image, Political Affiliation, Value of

Table 4.3 Relative importance index

Criteria	Weightage
Academic quality	0.96
Campus life	0.91
Health	0.88
Relationship with peers	0.85
Security	0.82
Sense of belonging	0.81
Social life	0.80
Living environment	0.80
School facilities/Services	0.79

Diversity, Development Opportunities, Time Management, Economic (Income and Job) and Leisure/Extracurricular Activities are of less importance and have therefore been removed from the analysis.

4.4.3 AHP survey

This survey aims to rank the criteria on the basis of their significance to the Subjective Happiness of students by using AHP pairwise comparison. AHP was adopted as the Multi-Criteria Decision-Making Methods (MCDM) method in our research to help rank the criteria. However, usually, AHP involves multiple decision-makers, each with their own opinions. Different decision-makers will have different opinions and views on the importance of the criteria and sub-criteria in the AHP model (Ossadnik et al., 2016). Consequently, it is of high importance to carefully aggregate the DM's opinions to come up with a single group judgment that can be used as input to the AHP model and achieve the goal of the analysis at hand (Yap et al., 2019).

For the context of this study, we adopted the Aggregation of Individual Judgments (AIJ) as shown in Equation (4.2), which is applied in the case of homogeneous groups where the judgment of all individuals has a similar priority.

$$a_{ij}^{[G]} = \prod_{k=1}^{r} 1\left(a_{ij}^{[k]}\right)_k^{\beta} \tag{4.2}$$

Note that: $a_{ij} = 1/a_{ji}$, for $i \neq j$, and $a_{ii} = 1$, all i.

Let us suppose n alternatives $(A_i, i = 1, ..., n)$ and r number of decision-makers $(D_k, k = 1, ..., r)$, let $A^{[k]} = (a_{ij}^{[k]})$ be the judgement matrix provided by the k-th decision-maker when comparing n elements $(i, j = 1, ..., n)$ and let β_k be the weight that the k-th decision-maker $(k = 1, ..., r)$ has informing the group decision. In this case, the weight of k-th decision-maker is considered equal. Thus, the study adopts Equation (4.2) to aggregate the individuals' judgements to be used for the AHP analysis as shown in Table 4.4.

Table 4.4 Aggregation of individual judgement using geometric mean

Criteria	CL	AQ	SF	SL	RWP	S	H	LE	SOB
Campus life (CL)	1.00	0.34	0.65	0.49	0.82	0.27	0.19	0.39	0.46
Academic quality (AQ)	2.95	1.00	2.73	2.28	1.75	0.82	0.84	1.63	3.20
School facilities/services (SF)	1.53	0.37	1.00	0.93	1.22	0.42	0.40	0.89	1.41
Social life (SL)	2.03	0.44	1.08	1.00	1.07	0.37	0.33	1.16	0.99
Relationship with Peers (RWP)	1.21	0.57	0.82	0.94	1.00	0.36	0.26	0.42	0.75
Security (S)	3.67	1.21	2.39	2.68	2.80	1.00	1.18	2.02	3.42
Health (H)	5.29	1.18	2.47	3.06	3.86	0.85	1.00	5.09	3.60
Living environment (LE)	2.54	0.61	1.13	0.86	2.39	0.57	0.20	1.00	1.96
Sense of belonging (SOB)	2.17	0.31	0.71	1.01	1.34	0.29	0.20	0.51	1.00

Table 4.5 Priority matrix

Criteria	Weightage	Rank
Campus life	0.44	8
Academic quality	0.61	7
School facilities/services	0.75	4
Social life	0.76	3
Relationship with peers	0.61	6
Security	0.93	2
Health	0.31	9
Living environment	0.96	1
Sense of belonging	0.63	5

The AHP analysis aids the decision-makers to take strategic decisions by allow-ing them to rank criteria in pairwise comparison. The AHP analysis thus calcu-lates the set of weighted criteria that can be adopted by the decision-makers based on their ranking and importance to the research as shown in Table 4.5.

It can be, therefore, concluded from the table that the students perceive Living Environment as the most important criteria to their Subjective Happiness in an academic environment. In addition, to ensure the consistency of judgment in pairwise comparison. The consistency Index and Consistency ratio for an AHP analysis can be calculated as shown in Equation (4.3):

$$CR = \frac{CI}{RI} \quad \text{and} \quad CI = \frac{\lambda_{max} - n}{n-1} \tag{4.3}$$

whereby, N is the number of criteria adopted for pairwise comparison which in our case is 9, and RI is the random index in our case is 1.45 as identified from the random index scale suggest by Saaty (1989) as shown in Figure 4.1.

In addition, Saaty suggests that if the consistency ratio exceeds 0.1 it indicates the set of judgment are inconsistent and not reliable.

In this study, the CR is 0.018, which is less than 0.1 as shown in Table 4.6; thus this indicates that the degree of consistency of judgment in AHP analysis is

Matrix size	Random consistency index (RI)
1	0.00
2	0.00
3	0.58
4	0.90
5	1.12
6	1.24
7	1.32
8	1.41
9	1.45
10	1.49

Figure 4.1 Saaty's Random Index Scale

Table 4.6 Consistency ratio

Consistency Index	0.026226		
RI	1.45		
CR	0.018	<	0.1

acceptable and reliable. Furthermore, to measure the consistency of two different forms of the test, this study adopted a parallel form (alternate form) of reliability.

4.4.4 Parallel form (alternate form) reliability

Parallel form or alternate form reliability is the measure of consistency for at least two unique types of test (Heale and Twycross, 2015). The simplest approach to measure reliability is to develop a larger set of similar questions, then randomly divide them into smaller tests. The reliability can be calculated by calculating the correlation between the sets of tests conducted on similar respondents or calculating the difference between the results. This study, therefore, adopts a parallel form or alternate form of reliability by using RII and AHP analysis to determine the student's perception of the importance of criteria that have an impact on their "Subjective Happiness" in an academic environment as shown in Table 4.7.

Thus, it can be seen from the table above that very little difference exists between the student's perception of the significance of criteria by adopting a different form of tests such as RII and AHP. Therefore, it can be concluded that parallel form or alternate form reliability exists, and the results are consistent. However, any difference above 0.5 and closer to 1 will be considered unacceptable.

4.5 Conclusion

Happiness is considered a form of social capital, which has increased the tangible benefits for society. As such communities that are rich in social capital are less likely to spend money and time in hospitals, prisons and depressions centres.

Table 4.7 Parallel form (alternate form) reliability

Criteria	RII weightage	AHP weightage	Difference
Academic quality	0.96	0.61	0.35
Campus life	0.91	0.44	0.56
Health	0.88	0.31	0.57
Relationship with peers	0.85	0.61	0.24
Security	0.82	0.93	0.14
Sense of belonging	0.81	0.63	0.17
Social life	0.80	0.76	0.04
Living environment	0.80	0.96	0.16
School facilities/Services	0.79	0.75	0.04

Therefore, it is important for a community or country to determine the happiness and satisfaction level of its people. One of the most important aspects is the university, is for the institution to measure the happiness and satisfaction of its students to keep on improving. This study determined the students' perception of the importance of a set of underpinning criteria that contribute to their happiness in an academic environment by adopting a parallel form (alternate form) of reliability to measure the consistency of the findings. As such, this study applied RII and AHP analysis by targeting students from the American University of Sharjah in the United Arab Emirates. The findings suggest that there exists little difference between the students' perceptions of the happiness criteria through the adaptation of these parallel forms of reliability measures.

References

Alexandrova, A., 2005. Subjective well-being and Kahneman's 'objective happiness'. *Journal of Happiness Studies*, 6(3), pp. 301–324.

Applasamy, V., Gamboa, R.A., Al-Atabi, M. and Namasivayam, S., 2014. Measuring happiness in academic environment: A case study of the School of Engineering at Taylor's University (Malaysia). *Procedia-Social and Behavioral Sciences*, 123, pp. 106–112. doi: https://doi.org/10.1016/j.sbspro.2014.01.1403.

Chan, G., Miller, P.W. and Tcha, M., 2005. Happiness in university education. *International Review of Economics Education*, 4(1), pp. 20–45. doi: https://doi.org/10.1016/S1477-3880(15)30139-0.

Chow, H.P., 2005. Life satisfaction among university students in a Canadian prairie city: A multivariate analysis. *Social Indicators Research*, 70(2), pp. 139–150. doi: https://doi.org/10.1007/s11205-004-7526-0.

Cohen, S., Kamarck, T. and Mermelstein, R., 1983. Perceived stress scale (PSS). *Journal of Health and Social Behavior*, 24, p. 285.

Csikszentmihalyi, M., 2013. *Flow: The psychology of happiness*. Random House.

Crawford, J.R. and Henry, J.D., 2004. The Positive and Negative Affect Schedule (PANAS): Construct validity, measurement properties and normative data in a large non-clinical sample. *British Journal of Clinical Psychology*, 43(3), pp. 245–265.

Diener, E.D., Emmons, R.A., Larsen, R.J. and Griffin, S., 1985. The satisfaction with life scale. *Journal of Personality Assessment*, 49(1), pp. 71–75. doi: https://doi.org/10.1207/s15327752jpa4901_13

Easterlin, R.A., 1974. Does economic growth improve the human lot? Some empirical evidence. In *Nations and households in economic growth* (pp. 89–125). Academic Press. doi: https://doi.org/10.1016/B978-0-12-205050-3.50008-7

Extremera, N. and Fernández-Berrocal, P., 2014. The subjective happiness scale: Translation and preliminary psychometric evaluation of a Spanish version. *Social Indicators Research*, 119(1), pp. 473–481. doi: https://doi.org/10.1007/s11205-013-0497-2

Frey, B.S. and Stutzer, A., 2006. *Should we maximize national happiness?* Institute for Empirical Research in Economics. Conference on New Directions in the Study of Happiness October 22–24, 2006, University of Notre Dame.

Heale, R. and Twycross, A., "Validity and Reliability in Quantitative Qtudies", *Evidence-based Nursing*, 18(3), 66–67.

Kammann, R. and Flett, R., 1983. Affectometer 2: A scale to measure current level of general happiness. *Australian Journal of Psychology*, 35(2), pp. 259–265.

Kiamarsi, A., Momeni, S, 2013. 'A survey of the relationship between social capital and happiness and academic achievement in female students', *Journal of School Psychology*, 2(1), pp. 119–130.

Lyubomirsky, S. and Lepper, H.S., "A Measure of Subjective Happiness: Preliminary Reliability and Construct Validation," *Social Indicators Research*, 46(2), pp. 137–155, 1999, doi: https://doi.org/10.1023/A:1006824100041.

Lyubomirsky, S. and Boehm, J.K., 2010. Human motives, happiness, and the puzzle of parenthood: Commentary on Kenrick et al. *Perspectives on Psychological Science*, 5(3), pp. 327–334. doi: https://doi.org/10.1177/1745691610369473

Medvedev, O.N. and Landhuis, C.E., 2018. Exploring constructs of well-being, happiness and quality of life. *PeerJ*, 6, p. e4903. doi: https://doi.org/10.7717/peerj.4903

Mordor Intelligence, 2020. "United Arab Emirates Cosmetic Products Market - Growth, Trends, Covid-19 Impact, and Forecasts (2021–2026)". Available at https://www.mordorintelligence.com/industry-reports/united-arab-emirates-cosmetics-products-market-industry

Olanipekun, A. O., Xia, B., Hon, C. and Darko, A., 2018. Effect of motivation and owner commitment on the delivery performance of green building projects. *Journal of Management in Engineering*, 34(1), 04017039.

Ossadnik, W.; Schinke, S. and Kaspar, R.H. 2016. Group aggregation techniques for analytic hierarchy process and analytic network process: A comparative analysis, *Group Decision and Negotiation*, 25(2), pp. 421–457. ISSN 1572-9907. doi: https://doi.org/10.1007/s10726-015-9448-4

Saaty, T.L., 1989. Group decision making and the AHP. In *The analytic hierarchy process* (pp. 59–67). Springer, Berlin, Heidelberg.

Scitovsky, T, 1976. *The joyless economy: An inquiry into human satisfaction and consumer dissatisfaction*. Oxford University Press.

Tabbodi, M., Rahgozar, H. and Makki Abadi, M.M., 2015. "The Relationship between Happiness and Academic Achievements", *European Online Journal of Natural and Social Sciences: Proceedings*, vol. 4, no. 1, p. 241, 2015.

Yap, J.Y.L.; Ho, C.C.; Ting, C.-Y., 2019. *Aggregating Multiple Decision Makers' Judgement* (pp.13–21). Springer, Singapore.

Yasin. M.A.S. Md and Dzulkifli. M, "The Relationship between Social Support and Psychological Problems among Students," 2010.

Zhang, J.M., Lu, J. and Sui, W.J., 2012. Campus planning and design research based on college students' happiness rating-case study of Shandong Jianzhu University Campus. *Advanced Materials Research*, 450, pp. 1123–1127. doi: https://doi.org/10.4028/www.scientific.net/AMR.450-451.1123

Zulkifli, I., 2013. Happiness and Students' Performance in Quantitative Subjects–A Preliminary Study. *Prosiding Book of ICEFMO*.

Part III

Validity test: Research case study examples

5 Enhancing reliability and validity in a study exploring the indicators of a sustainability assessment framework for neighbourhood development in Nigeria

Ayomikun Solomon Adewumi, Dumiso Moyo and Vincent Onyango

5.1 Background

In the last 100 years, urbanization has been more resource-demanding, which amongst other things, contributes significantly to climate change, loss of soil carbon, deforestation, loss of biodiversity and adverse effects on the living standards of people (Owen, 2010; Komolafe et al., 2014). As more people will be living in cities with more pressure on the built environment (UN-Habitat 2015a), it has been argued that humanity will fail or succeed in the battle for sustainability in urban areas (Girardet, 2008, 2015; Gehl, 2010; UN-Habitat, 2015b). It is noteworthy that the urgency of sustainability in the urban areas has been underpinned in the international discourse and political agreements by force-acting and motivating decisions. These include new urban agenda; sustainable development goals (SDGs) 11 aimed at delivering sustainable communities; and recently the new urban sustainability frameworks of 2018.

However, the slow progress recorded in the campaign for urban sustainability (Berardi, 2011; Huang et al., 2016) necessitated the need for a re-examination of the processes and methods currently being adopted to accommodate urban growth (Smith, 2015; UN-Habitat, 2016). It originates with the idea that urban places are a product of a decision-making process and that, whether an urban area will be sustainable or not depends on the integration and consideration of sustainability indicators (SIs) in the decision-making process. This understanding of the role of the decision-making process to deliver sustainable urban places heralded an approach that has increasingly been tried, using the neighbourhood as a component building block (Wangel et al., 2016; Berardi, 2013). The main argument is that if SIs are integrated at the decision-making process of a new neighbourhood, then this can in the wider picture create a sustainable urban area (Cole, 1999; Komeily and Srinivasan, 2015). SI is traceable to the Rio Summit of 1992 as captioned in chapter 40 of Agenda 21 on the need to establish "indicators of sustainable development" which will assist in monitoring progress (Bell and Morse, 2008).

DOI: 10.1201/9780429243226-8

This led to the development of neighbourhood sustainability assessment frameworks in the closing decades of the 20th century through which a proposed neighbourhood can be assessed against an array of SIs (Wangel et al., 2016), and therefore enhancing decisions about the sustainable designs and options that can be applied in that context.

Whilst the use and further development of these assessment frameworks continue in the developed nations (Joss et al., 2015; Sharifi and Murayama, 2015) with the development of BREEAM Communities, LEED ND and green star communities amongst others, ostensibly to deliver more sustainable places (USGBC, 2016), most Sub-Sahara African (SSA) countries including Nigeria are yet to evolve such frameworks for decision-making to deliver sustainable urban places. Nigeria is of interest as within a period of 30 years (1952–1982), major cities like Lagos, Kano, Port Harcourt, Maiduguri, Kaduna, Jos and Ilorin have experienced a five-fold increase in population (Onibokun and Faniran, 1995). In 2025, it is projected that 60% of the Nigerian population will be living in urban centres (FMLHUD, 2014).

To this end, the study is underpinned by the problem that this growing urban population in Nigeria calls for an approach and mechanism for decision-making that will enhance planning for such growth. In the absence of which, there would be little certainty on how to trade-off and balance competing interests and ideas; and how to calibrate the aspirations, means and delivery mechanisms with the futuristic solutions that are sustainable given the implications of urbanization in Nigeria. Also, given that decision-making is at the heart of planning, it is feared that lack of evidence-driven decision-making approaches can lead to unsustainable urban development in Nigeria. Moreover, current assessment frameworks for decision-making developed in the western countries are "tailor-made" and context-specific (Komeily and Srinivasan, 2015; Wangel et al., 2016; Bina, 2008), making the idea of adopting any of these frameworks for use in another context, like Nigeria, challenging and problematic.

The study aims to explore the indicators of a sustainability assessment framework that can be used in the decision-making process in the development of new neighbourhoods in metropolitan Lagos, Nigeria. These three key objectives were pursued: (i) Distil the SIs for assessing new neighbourhood in metropolitan Lagos based on stakeholders' perception; (ii) Rank the profile of stakeholders' preferences of the distilled SIs through the analytical hierarchy process (AHP); (iii) Validate the ranked indicators by testing their potential on usability, and adoptability in the context of metropolitan Lagos.

The next section presents the methodology adopted for the study as informed by the study aim and objectives. The validity and reliability of the research design which this chapter set to deliver were explored using the following key questions: (i) How acceptable and useful are the findings of the study by key institutions in the decision-making process of a new neighbourhood? (ii) How consistent and representative are the data obtained? (iii) What is the applicability of the findings in other context outside metropolitan Lagos? (iv) Could an independent researcher undertake the same study and obtain similar results, or analyse the data of the study and obtain similar findings?

5.2 Research methodology

5.2.1 Research design

The need to establish stakeholders' perceptions and preferences of SIs in a particular context which can be challenging, led to the choice of the critical realism philosophical stance for this study. In addition, critical realism offers an explanatory linkage that integrates the people's perceptions in relation to their context (Easterby-Smith et al., 2012). The study employed both inductive and deductive approach for its analysis. This involved generating new theory, and the statistical testing of theory focused on establishing statistically the importance of the pre-identified indicators from literature amongst a list of others. Also, a mixed method approach that applies both numerical and textual data as influenced by the research objectives was adopted for the study. As this study is focused on urban neighbourhoods in Nigeria, the intrinsic case study research strategy was adopted which would involve a detailed study of a case, which in this instance is metropolitan Lagos. This could perhaps be a fair representative for other cities in Nigeria. The study adopted questionnaire survey as a technique for data collection using a semi-structured type of questionnaire with both closed-ended and open-ended questions to capture both quantitative and qualitative data, respectively. The research design is summarized in Table 5.1.

5.2.2 Questionnaire structure, analysis and sampling technique

Two different questionnaires were administered. The first questionnaire sought to distil the indicators. Using a 5-point Likert scale (1- Not important and dispensable; 2- Little importance but contribute insignificantly; 3- Important but only contributes slightly; 4- Important and contributes significantly; 5- Highly important and indispensable), respondents' perceptions were captured on the importance of the indicators in contributing to sustainable neighbourhood in metropolitan Lagos. The results were analysed using descriptive statistics by calculating the weighted averages (WA), co-efficient of variation (CV) and

Table 5.1 Summary of research design

Research design	Choice	Justification
Philosophy	Critical realism	To interpret the findings in relation to the context
Approach	Inductive and deductive	To generate and test theory
Methods	Mixed (quantitative and qualitative)	Numeric and textual data are required
Strategy	Case study	Allows for in-depth study and exploration of a single case
Tools	Questionnaire	To capture stakeholders' perceptions of sustainable neighbourhoods and its indicators

content validity ratio (CVR). An indicator with a CV value less than 0.5 can be said to be consensually agreed upon by the stakeholders (Wilson et al., 2012; Lawshe, 1975). Similarly, is the CVR developed by Lawshe (1975) which measures agreement in a survey in order to know how essential a particular item is to the respondents. According to Lawshe (1975), the CVR values ranges from –1 (perfect disagreement) to +1 (perfect agreement). Values above zero indicates that over half of the respondents agree that a variable is essential (Ayre and Scally, 2014). It can be deduced that an indicator with a CVR value equal or greater than 0.29 is essential based on stakeholders' perception (Wilson et al., 2012; Lawshe, 1975).

The questionnaire was administered to the two main categories of stakeholders:

1   Institutional stakeholders which are of three sub-categories namely: developers (private or government); regulators; and other built environment professionals. Participants identified in the category of the "regulators" and "developers" were nominated by the each of the institutions responsible for neighbourhood development. Also, two private real estate companies were contacted to make nominations. Participants who are built environment professionals and academic were identified using the snow-ball and random sampling, respectively. It is noteworthy that the sampling technique adopted for the institutional stakeholders was helpful to achieve a high response (see Table 5.2).

2   Residents who are consumers of the neighbourhoods. As this study could not possibly engage all neighbourhoods in metropolitan Lagos due to resources and time, three neighbourhoods that are of the master planned neighbourhood development in metropolitan Lagos were identified. Two from neighbourhood development delivered by Government institutions. These are Abesan Estate delivered by the State government (Neighbourhood A) and Gowon Estate developed by the Federal government (Neighbourhood B). One neighbourhood named Rose Garden Estate delivered by a private institution (Neighbourhood C) was further identified. The questionnaire was administered in these neighbourhoods using the purposive sampling. This is a technique whereby the researcher relies on his judgement when choosing members. The purposive sampling was adopted in: (i) neighbourhood A using the design typologies; (ii) neighbourhood B using sectoral divisions; and (iii) neighbourhood C by administering two questionnaires in each of the blocks which comprise of 4 flats each that make up the neighbourhood. In all 309 questionnaires were retrieved with a response rate of 81.97%, 70.2% and 71.43% in neighbourhoods A, B and C, respectively.

The second questionnaire sought to capture stakeholders' preferences of the distilled indicators using the AHP which would further help to determine and confirm the rank and priorities of the indicators in the decision-making process of a new neighbourhood. Participants were first asked to compare in pairs the dimensions (environmental, socio-cultural and economic) with each other, and later the indicators under each sustainability dimension. For instance, they were required

Table 5.2 Questionnaire distribution for institutional stakeholders' perception of indicators

Institutional stakeholders	Questionnaires administered	Questionnaires retrieved
Regulators		
Ministry of physical planning and urban development (MPPUD)	1	1
Lagos state building control agency (LASBCA)	1	1
Lagos state physical planning and development authority (LASPPDA)	1	1
New town development authority (NTDA)	1	1
Developers		
Ministry of Housing (MoH)	1	1
Lagos state development and property corporation (LSDPC)	1	1
Lagos building investment company (LBIC)	1	1
Private developer (Jubilee homes)-PDEV1	1	1
Private developer (Jide Taiwo and co)-PDEV2	1	1
Built environment professionals (individuals)		
New town development authority	3	3
Ministry of Housing	9	4
Lagos building control agency	3	2
Academics		
Obafemi Awolowo University, Nigeria	2	2
University of New South Wales, Australia	1	1
Total	27	21
Response rate	75%	

to respond to a question such as: "How important is environmental dimension relative to economic dimension in contributing to a sustainable neighbourhood?" This preference was obtained using the 9-point scale between 1 (representing equal importance) and 9 (representing extreme importance). However, the reciprocal of value was assigned to the other dimension or indicator that is paired with.

The participants for this phase were those who volunteered that they were still interested in the next stage after completing the first questionnaire. However, new participants were also included in the pair-wise comparison to achieve an acceptable sample size (see Table 5.3).

5.3 Results and main findings

5.3.1 Perception and preferences for indicators

Using Microsoft excel functions, the WA; standard deviation (SD); CV; and the CVR of the indicators were calculated (see Table 5.4) based on the perceptions from both categories of stakeholders. The two indicators that did not reach consensus based on institutional stakeholders perceptions are home garden to support local food production and active frontage to support shops because they both

Table 5.3 Questionnaire distribution for stakeholders' preferences

Category of respondents	Questionnaire administered	Questionnaire retrieved	Number of valid questionnaire
Residents			
Neighbourhood A	5	2	1
Neighbourhood B	5	3	2
Neighbourhood C	5	1	1
Private developers	5	4	3
Ministry of Housing			
• Architecture and building services	5	5	4
• Quantity surveying	5	2	1
• Engineering	5	2	2
• Town planning	5	5	4
Lagos state property development corporation	5	2	2
Lagos state building investment company	5	3	1
Total	50	29	21

have CVs greater than 0.5 and CVRs less than 0.29. The result was similar to the one from residents' perception. None of the 25 pre-selected indicators had a CV less than 0.5 but three have a CVR less than 0.29 (see Table 5.4). Two of which are "home garden to support local food production" and "active frontage to support shops". The other one is "use of locally made material" although with the lowest CVR but with highest CV of 0.44 will be part of the distilled indicators because of a higher rating average of 4.00 it received from the institutional stakeholders.

Having distilled the indicator set based on stakeholders' perception, Table 5.5 presents the result from the stakeholders' preferences of the indicators under their respective dimensions of sustainability (i.e. environmental, socio-cultural and economic) using the AHP from which their local and global priority values of the indicators were calculated which helped in ranking the indicators. The aggregate values obtained for each indicator is known as the local priority value, which is the weight of the indicator when compared with other indicators under their respective dimensions. However, there is need to calculate the global priority value which shows the weight of an indicator when compared with other indicators. This was calculated by multiplying the local priority value and the weight of the dimension where it belongs. For example, renewable energy with a local priority value of 0.89 has a global priority value of 0.037 (that is, 0.98 multiply by 0.379).

The cost of construction, operation and maintenance was ranked first in the indicator set while neighbourhood square was ranked lowest. The weighing and ranking of the indicators showed a fair distribution across the dimensions of sustainability. Environmental was allocated 0.379; economic-0.311 and socio-cultural-0.310. The sustainability index which forms a key part of the assessment framework is to guide in the decision-making process of a new

Table 5.4 The rating averages, CV and CVR of the indicators based on stakeholders perception (n = 21 for institutional stakeholders; n = 309 for residents)

Indicators	WA		SD		CV		CVR	
	Inst.	Res.	Inst.	Res.	Inst.	Res.	Inst.	Res.
Use of renewable energy	4.29	4.01	1.44	1.34	0.34	0.33	0.71	0.50
Waste collection and management	4.76	4.43	1.72	1.51	0.36	0.34	1.00	0.80
Facility management	4.57	4.24	1.59	1.42	0.35	0.34	0.90	0.72
Environmental Impact Assessment	4.24	3.97	1.42	1.33	0.34	0.33	0.52	0.46
Pollution control	4.38	4.34	1.49	1.47	0.34	0.34	0.81	0.78
Green field preservation	4.10	3.76	1.37	1.29	0.33	0.34	0.52	0.31
Effective land usage	4.00	3.81	1.34	1.30	0.33	0.34	0.52	0.39
Efficient use of resources	4.29	4.67	1.44	1.66	0.34	0.35	0.52	0.90
Outdoor spaces	4.57	4.33	1.59	1.46	0.35	0.34	0.71	0.65
Aesthetics	3.67	3.82	1.29	1.30	0.35	0.34	0.24	0.47
Quality of construction material	4.86	4.45	1.78	1.53	0.37	0.34	1.00	0.81
Good pedestrian lane	4.76	4.50	1.72	1.56	0.36	0.35	1.00	0.84
Diverse mobility option	4.19	4.24	1.40	1.42	0.33	0.34	0.71	0.66
Nearness to amenities & infrastructures	4.43	4.33	1.51	1.46	0.34	0.34	0.81	0.70
Availability of infrastructure & amenities	4.86	4.64	1.78	1.64	0.37	0.35	1.00	0.88
Security	4.62	4.43	1.62	1.51	0.35	0.34	0.81	0.77
Access to reliable and portable water	4.71	4.48	1.69	1.54	0.36	0.34	0.81	0.81
Inclusive design	4.05	4.13	1.35	1.38	0.33	0.33	0.43	0.62
Use of locally made material	4.00	3.18	1.34	0.44	0.33	0.44	0.43	-0.08
Provision of neighbourhood square	4.19	3.91	1.40	0.34	0.33	0.34	0.71	0.39
Home affordability	4.24	4.39	1.42	0.34	0.34	0.34	0.52	0.79
Support for home-based business	3.62	4.22	1.29	0.33	0.36	0.33	0.34	0.71
Cost of construction, operation, & maintenance	4.43	4.40	1.51	0.34	0.34	0.34	0.81	0.83
Home garden for local food production	**3.00**	**3.54**	**1.60**	**1.30**	**0.53**	**0.37**	**-0.24**	**0.19**
Active frontages to encourage shops	**2.38**	**3.20**	**1.85**	**1.38**	**0.78**	**0.43**	**-0.71**	**0.00**

Table 5.5 Weight and ranking of the distilled indicators based on stakeholders' preferences

Dimensions	Indicators	Local priority	Global priority	rank
Environmental (0.379)	Environmental impact assessment	0.169	0.064	4
	Efficient use of resources	0.158	0.060	5
	Pollution control	0.135	0.051	6
	Waste collection and management	0.128	0.049	7
	Strategy to maintain infrastructure	0.116	0.044	8
	Effective land usage	0.107	0.040	9
	Use of renewable energy	0.098	0.037	10
	Green field preservation	0.090	0.034	11
Social-cultural (0.310)	Access to portable water	0.116	0.036	12
	Availability of infrastructure & amenities	0.113	0.035	13
	Quality of construction material	0.110	0.034	14
	Security	0.100	0.031	15
	Nearness to basic amenities	0.094	0.029	16
	Use of locally made material	0.081	0.025	17
	Outdoor spaces	0.071	0.022	18
	Diverse mobility option	0.071	0.022	18
	Inclusive design	0.065	0.020	20
	Aesthetics	0.061	0.019	21
	Good pedestrian lane	0.061	0.019	21
	Neighbourhood squares	0.058	0.018	23
Economic (0.311)	Cost of construction of operation, & maintenance	0.398	0.124	1
	Home affordability	0.324	0.100	2
	Support for home-based business	0.278	0.087	3
			1.00	

neighbourhood how indicators should be prioritized while also allowing a proposed neighbourhood to be assessed and scored using the sustainability index. Prior to the actual construction of a neighbourhood, it can be assessed on a scale of 0 to 1. Using the sustainability index for example in the decision-making process of a new neighbourhood can create the following scenarios: One, meeting the assessment criteria of all the environmental indicators would result to a score of 0.379 on a scale of 0 to 1, which is relatively low. Two, consideration for only the assessment criteria in both environmental and economic dimensions would result to a score of 0.690. Three, to achieve a score of 0.80 on a scale of 0 to 1, it will require adequate consideration across the three dimensions in the sustainability index.

5.4 Validity and reliability

Validity is "the property of a research instrument that measures its relevance, precision and accuracy" (Sarantakos, 2013:99). It is the verification process of the findings in a research showing whether the research measures what it was intended to measure and how valid or truthful the research findings are (Dangana, 2015; Sarantakos, 2013). Although validity instruments used for qualitative research differ from that of quantitative research, both serve the purpose of checking the quality of data, results and interpretation (Creswell and Plano Clark, 2011). Besides validation, this study also examined the reliability of the research design which is "concerned with the question of the extent to which one's findings will be found again" (Merriam 1995:55). It confirms if the same result would be obtained using similar data collection procedure (Easterby-Smith et al., 2012; Sarantakos, 2013:104). To this end, the following were addressed to ascertain the validity and reliability of the research design.

5.4.1 Internal validity

This measures the truthfulness of the research process and findings using appropriate criteria (Merriam, 1995). It raises the question of the consistency of the data obtained. In this study, statistical tests served as the instrument to confirm the levels of robustness of the quantitative data. The study did not only rely on the CV as the status of each of the indicators were further confirmed by their CVR values in a process known as triangulation. The internal validity of the study was also strengthened on the basis that the results are consistent with existing studies in what Sarantakos (2013), Denzin (1970) and Mathison (1988) referred to as cumulative internal validity. Cost of construction, operation and maintenance which was highly ranked based on stakeholders' preferences supports Ijasan and Ogunro (2014) and Ibem et al. (2013) who advocated for affordable maintenance system for urban neighbourhoods. This was also considered in the pearl community rating system (PCRS) which allocated 2.5% of its total weighing to life cycle costing (IDP-1). In addition, home affordability which was ranked second concretizes (Olotuah and Aiyetan 2006 and Hamiduddin 2015) that emphasize affordability as crucial in delivering sustainable neighbourhood in metropolitan Lagos.

It is also noteworthy that the findings showed some similarities with some assessment frameworks in terms of preference when the indicators are compared with each another.

The findings agree with:

1 BREEAM communities that:

   • Waste collection and management has priority over strategy to maintain infrastructure;
   • Strategy to control pollution has priority over waste management;
   • Environmental impact assessment has priority over efficient use of resources; and effective land usage;
   • Social amenities and infrastructure have priority over security of lives and properties (BRE, 2012).

2 Pearl community rating system (PCRS) that:

   • Cost of construction, operation and maintenance has priority over home affordability;
   • Social amenities and infrastructure have priority over security of lives and properties (AUPC, 2010).

Table 5.6 Respondents' affiliation and roles in neighbourhood planning

Institutions and pseudonyms	Role
Ministry of Works (MOW)	Developer (Govt)
Ministry of Housing (MOH_1)	Developer (Govt)
New town development authority (NTDA)	Regulator
Ministry of physical planning and urban development (MPPUD)	Regulator
Lagos state property and development corporation (LSDPC)	Developer (Govt)
Private developer (PDEV_1)	Developer
Private developer (PDEV_2)	Developer
Lagos building investment company (LBIC)	Regulator
Lagos building control agency (LABCA)	Regulator

3   Green star communities that:

- Home affordability has priority over support for home-based business;
- Nearness to basic amenities has priority over diverse mobility options;
- Social amenities and infrastructure have priority over security of lives and properties (GBCA, 2012).

The internal validity was further strengthened by presenting the results to stakeholders to ascertain the acceptability and usefulness of the findings (see Table 5.6). Using a 5 -point Likert scale (1 - strongly disagree; 2 - disagree; 3 - neutral; 4 - agree; 5 - strongly disagree), participants were asked to indicate their level of agreement on the following: (i) the comprehensiveness of the indicators in addressing sustainability issues; (ii) ranking of indicators; and (iii) adoptability of the indicators.

Aggregate results (see Table 5.7) showed that all the institutions agreed on the comprehensiveness, ranking, usability and adoptability of the indicator set in addressing sustainability at the neighbourhood level in metropolitan Lagos. Each of the parameter for assessment had a WA greater than 4 indicating a high level of agreement.

Table 5.7 Content validity of the indicator set

Institutions	Level of agreement (1 - strongly disagree; 2 - disagree; 3 - neutral; 4 - agree; 5 - strongly agree)		
	Comprehensiveness	Ranking of indicators	Adoptability
LSDPC	5	4	4
MOW	5	5	5
NTDA	5	4	5
LBCA	5	5	4
MPPUD	5	5	5
MOH_1	4	4	4
PDEV_1	4	4	4
LBIC	4	4	4
PDEV_2	5	5	4
WA	4.67	4.44	4.33

Explaining further their judgement on the adoptability and uptake of the SIs for use in their various institutions in the decision-making process of a new neighbourhood, MOH_1 noted that "the development of sustainable cities and communities is one of the SDGs to which Nigeria is a signatory". MOW_1 corroborating this position said that "using the indicators in decision-making would ensure the delivery of quality housing to the end-users". NTDA_1 noted that "the indicators are strongly essential in decision-making for a new neighbourhood because they help to better design a functional neighbourhood and livelihood enhancing factors". PDEV_1 agreed on the basis that its institution is "receptive to whatever will enhance the goal of affordable housing delivery both in quantity and quality which the indicator epitomizes". LBIC posited that "if the aforesaid indicators are successfully put to use, a sustainable neighbourhood would be built, which would enhance the lives and properties of people".

5.4.2 External validity

This refers to the extent in which the research findings can be generalized to a wider population (Dangana, 2015). It shows the conviction that the research findings are applicable to a large population. In this study, it addresses the question of the representativeness of the data obtained. A sample size of 309 residents seems to be representative across the various residents of neighbourhoods in metropolitan Lagos. This is also because the 309 respondents are of different social characteristics as captured by the questionnaire. While the representativeness of the sample size of 21 to capture stakeholders' preferences of the indicators may be questioned, it is noteworthy, that the AHP does not rely on large sample size for validity as it is applied on a research focusing on a specific issue (Cheng et al., 2002; Schmidt et al., 2015). In addition, the AHP technique may be impossible (impracticable) for a survey with a large sample size as uninterested participants have a great tendency to provide arbitrary answers resulting to a high degree of inconsistency (Cheng et al., 2002). To this end, AHP has been conducted with small sample size in contrary to the traditional survey where a large sample size is required. For instance, Akadiri (2011), Cheng et al. (2002) and Dangana (2015) used 19; 9 participants; and 15 participants, respectively. In this study, participants were identified to cut across the categories of stakeholders.

In addition, while this study appreciates the peculiarities in various contexts about the perception of SIs, it is noteworthy that cities tend to encounter similar challenges of urbanization. To this end, the findings from this study can perhaps be generalized to other growing cities. This was already noticeable as some of the findings is like that obtained in western countries.

5.4.3 Internal reliability

This assures that the same result would be obtained if the research is repeated in the same context. This raises the question of whether an independent researcher in the same context can undertake the same study and obtain similar results, or analyse the data of the study and obtain similar findings. This was strengthened in this study through the transparency of the research design and process. What could also enhance this, is the identification of the research participants in the same context that would likely result to similar opinions and perceptions based on values and culture. This for instance was justified in this study as the critical realism philosophical lens showed that stakeholders' perceptions and preferences of SIs were influenced by the current challenges of urbanization in metropolitan Lagos. To this end, stakeholders' perceptions and preferences in a repeated study are not likely to change so far, the urbanization challenges persist.

5.4.4 External reliability

This assures the same result if the study is to be repeated in another context using the same procedure. As this study is anchored on the context-specificity of sustainability, the result may be obtained in other SSA cities that have similar development trend, values and approach to physical planning. However, findings from this study have also established that there may perhaps be some similarities if the study is conducted in another context in terms of the meaning of a sustainable neighbourhood and the weighting of its indicators.

5.5 Main findings and conclusions

This chapter set out to address the validity and reliability of a research design formulated to explore the indicators of a sustainability assessment framework for neighbourhood development in Nigeria. This chapter demonstrated internal validity as the main results of the study were presented to stakeholders who agreed to the comprehensiveness of the indicators in addressing sustainability issues in metropolitan Lagos. More also, there was a high level of agreement on the ranking of the indicators and subsequent adoption in the decision-making process of a new neighbourhood. Furthermore, the results obtained from the stakeholders' perception which helped to distil the indicators were validated through using both their CV and CVR values. The CVR for instance was used to support the CV values through which the indicators were distilled based on the level of agreement amongst respondents. The reliability and validity also appreciate the context-specific nature of sustainability- a situation which might perhaps may result to obtaining different result in another context. This study however through critical realism philosophical lens which argues that people's perception and understanding are shaped by their experience could conclude that internal reliability was strengthened as decision-makers and residents are confronted with similar urbanization challenges.

   To this end, the study can arrive at the following conclusions: One, internal validity of a built environment research can best be enhanced by ensuring the acceptability and usefulness of the main findings. This is needful so that such study could make necessary impacts on the environment. Failure to this means that such research findings may only be an academic exercise with no policy implication that has the potential effect the desired change. Two, external validity which is the representativeness of a research design of a wider population can be strengthened using a justifiable sample size. However, researchers should be aware that some techniques like the AHP as adopted in this study to capture stakeholders' preferences do not require a large sample size. This is to reduce cases of inconsistencies attributed to large sample size. Also, some studies may require only the voices of experts which in that instance will be representative and be applicable to wider population.

Three, internal reliability is best enhanced through the transparency of the research design. For example, there is need for clarity of research philosophy which is the lens for the interpretation of the research findings. In this study, the critical realism helped to know that the perceptions and preferences of the stakeholders are shaped by the underlying reality. Similar results can be interpreted in several ways through the lenses of other philosophical positions such as pragmatism, post-modernism, or positivism amongst others. Four, although the role of context is significant in built-environment research, the same results are likely to be obtained in other cities due to similar challenges that confront the 21st-century cities.

References

Akadiri, O., 2011. Development of a multi-criteria approach for the selection of sustainable materials for building projects, 'Unpublished PhD thesis'. Wolverhampton: University of Wolverhampton. Retrieved from https://pdfs.semanticscholar.org/19b3/3292478137f4ccb7e9b3eea1ddf6235eab5f.pdf

AUPC, 2010. *The Pearl Rating System for Estidama.* Abu Dhabi: Abu Dhabi Urban Planning Council.

Ayre, C., & Scally, A. J., 2014. Critical Values for Lawshe's Content Validity Ratio: Revisiting the Original Methods of Calculation. *Measurement and Evaluation in Counseling and Development, 47*(1), pp. 79–86. doi:https://doi.org/10.1177/0748175613513808

Bell, S. & Morse, S., 2008. *Sustainability Indicators-Measuring the Immeasurable?* 2nd ed. London: Earthscan.

Berardi, U., 2011. Beyond sustainability assessment: Upgrading topics by enlarging the scale of assessment. *International Journal of Sustainable Building Technology and Urban Development, 2,* pp. 276–282.

Berardi, U., 2013. Sustainability assessment of urban communities through rating systems. *Environment Development and Sustainability, 15,* pp. 1573–1591.

Bina, O., 2008. Context and systems: Thinking more broadly about effectiveness in strategic environmental assessment in China. *Environmental Management, 42*(4), pp. 717–733.

BRE, 2012. *BREEAM Communities: Technical Manual SD202-01-2012.* Watford: Building Research Establishment.

Cheng, E. W., Li, H., & Ho., 2002. Analytic hierarchy process (AHP): A defective tool when used improperly. *Measuring Business Excellence, 6*(4), pp. 33–37. doi:https://doi.org/10.1108/13683040210451697

Cole, R., 1999. Building environmental assessment methods: Clarifying intentions. *Building Research and Information, 27,* pp. 230–246.

Creswell, J. & Plano Clark, V., 2011. *Designing and Conducting Mixed Methods Research.* Thousand Oaks, CA: Sage.

Dangana, S., 2015. A decision support framework for selecting innovative sustainable technologies for delivering low carbon retail buildings, Plymouth: Unpublished Thesis; Doctor of Philosophy.

Denzin, N. K., 1970. *The Research Act: A Theoretical Introduction to Sociological Methods.* Chicago: Aldine.

Easterby-Smith, M., Thorpe, R. & Jackson, P., 2012. *Management Research.* London: Sage Publications.

GBCA, 2012. *Green Star Communities: Guide for Local Government.* Melbourne: Green Building Council Australia.

Gehl, J., 2010. *Cities for People.* Washington: Island Press.

Girardet, H., 2015. "Ecopolis"-The regenerative city. In *Low Carbon Cities* (pp. 59–74). Oxon: Routledge.

Girardet, H., 2008. *Cities People Planet.* 2nd ed. Hoboken New Jersey: Wiley.

Hamiduddin, I., 2015. Social sustainability, residential design and demographic. *Town Planning Review*, 86(1), pp. 29–52.

Huang, L., Yan, L. & Wu, J., 2016. Assessing urban sustainability of Chinese megacities: 35 years after the economic reform and open-door policy. *Landscape and Urban Planning*, 145, pp. 57–70.

Ibem, E., Opoko, A., Adeboye, A. & Amole, D., 2013. Performance evaluation of residential buildings in public housing estates in Ogun State, Nigeria: Users' satisfaction perspective. *Frontiers of Architectural Research*, 2, pp. 178–190.

Ijasan, K. C. & Ogunro, V. O., 2014. How rapid urbanisation, neighbourhood management affects living conditions. In *A Survey of Agege Local Government Area.* Lagos, Nigeria, s.l.: Canada Center of Science and Education.

Joss, S. et al., 2015. *Tomorrow's City Today: Prospects for Standardising Sustainable Urban Development.* London: University of Westminster.

Komeily, S. & Srinivasan, R., 2015. A need for a balanced approach to neighbourhood sustainability assessment. *Journal of Sustainable Cities and Society*, 18, pp. 32–43.

Komolafe, A., Adegboyega, S., Anifowose, A., Akinluyi, F., & Awoniran, D., 2014. Air pollution and climate change in Lagos, Nigeria: Needs for proactive approaches to risk management and adaptation. *American Journal of Environmental Sciences*, 10(4), pp. 412–423.

Lawshe, C. H., 1975. A quantitative approach to content validity. *Personnel Psychology*, 28, pp. 563–575.

Mathison, S., 1988. Why triangulate? *Educational Researcher*, 17(2), pp. 13–17.

Merriam, S. B., 1995. What can you tell from an N of 1: Issues of validity and reliability in qualitative research. *Journal of Lifelong Learning*, 4, pp. 51–60.

Olotuah, A. & Aiyetan, A., 2006. Sustainable low-cost housing provision in Nigeria: A bottom-up participatory approach. In *Procs 22nd Annual ARCOM Conference*, 4–6 September (pp. 633–639). Birmingham: Association of Researchers in Construction Management.

Onibokun, A., & Faniran, A., 1995. *Urban Research in Nigeria.* Institut français de recherche en Afrique.

Owen, D., 2010. *Green Metropolis: Why Living Smaller, Living Closer, and Driving Less Are Keys to Sustainability.* New York: Riverhead Books.

Sarantakos, S., 2013. *Social Research.* 3rd ed. Hamisphire: Palgrave Macmillan.

Schmidt, K., Aumann, Hollander, I., Damm, K., & Matthias Graf von Schulenburg, J., 2015. Applying the Analytical Hierarchy Process in healthcare research: A systematic literature review and evaluation of reporting. *Medical Informatics and Decision Making*, 15(112). doi:10.1186/s12911-015-0234-7

Sharifi, A. & Murayama, A., 2015. Viability of using global standards for neighbourhood sustainability assessment: Insights from a comparative case study. *Journal of Environmental Planning*, 58(1), pp. 1–23.

Smith, M., 2015. Planning for urban sustainability: the geography of LEED-neighbourhood development (LEED-ND) projects in the United States. *International Journal of Urban Sustainable Development*, 7(1), pp. 15–32.

UN-Habitat, 2015a. A new strategy of sustainable neighbourhood planning. [Online] Available at: http:.//www.unhabitat.org [Accessed 02 03 2016].

UN-Habitat, 2015b. *Sustainable Urban Development in Africa*. Nairobi: United Nations Human Settlement Programme.

UN-Habitat, 2016. *Urbanization and Development: Emerging futures-WORLD CITIES REPORT 2016*. Nairobi: United Nations Human Settlements Programme.

USGBC, 2016. *LEED ND V4 for Neighbourhood Development*. Washington: US Green Building Council.

Wangel, J., Wallhagen, M., Malmqvist, T. & Finnveden, G., 2016. Certification systems for sustainable neighbourhoods: What do they really certify? *Environmental Impact Assessment Review*, 56, pp. 200–2013.

Wilson, F., Pan, W., & Schumsky, D., 2012. Recalculation of the critical values for Lawshe's content validity ratio. *Measurement and Evauation in Counselling and Development*, 45, pp. 197–210. doi:https://doi.org/10.1177/0748175612440286

6 A predictive validity analysis of water demand forecasting model in the UAE

Vian Ahmed, Sara Saboor, Ahmad Saad,
Hasan Saleh, Nikita Kasianov and Tahani Alnaqbi

6.1 Introduction

Over the past few years, climate change has become the biggest threat to the world causing an irreversible change in the ecosystem such as rises in sea level, food-insecurity, natural resources scarcity, seasonal disorders and an increase in natural disasters such as wildfires, flooding and droughts (Al-Qawasmi et al., 2019). However, among all the problems caused by climate change, water scarcity is believed to be the major troubling factor.

According to Gosling and Arnell (2016), the global scale assessment of climate change on water scarcity using the water crowding index shows that by 2050, 0.5 to 3.1 billion people will be exposed to an increase in water scarcity. In addition, according to a report published by the United Nations in 2018, water scarcity claims to be a potential threat for every continent (Boretti and Rosa, 2019). The scarcity can be accrued to physical shortages of water or inadequate access to water resources due to a lack of infrastructure.

Thus, due to the growing concerns of water scarcity, United Nations (UN) recognizes water as the significant resource in underpinning equitable, stable and productive societies and ecosystems, which affects the entire development agenda. As such, "Water connects us all" under the "Water Goal" was adopted is one of the seventeen sustainable development goals (Goal 6) to ensure water security (Ait-Kadi, 2016). This implies that Water scarcity is a global issue, while sustainable water management and water demand forecasting are significant goals that must be incorporated into the political agenda by policy makers to ensure water stability and cost management.

A study by Alsharhan and Rizk (2020) reports that 73% of the global affected population inhabit the Asian continent, whilst the 13 most water-stressed countries are found in the Middle East alone. Amongst the Middle Eastern countries, UAE ranks tenth, with one of the highest per capita residential water consumption rates in the world. During the last decade and due to the development of oil sector, the Gulf region in general and the UAE particularly has gained enormous attention from investors, which has shaped once the barren desert into a modern city, with the UAE government investing billions of dollars every year to improve the infrastructure of the country. The construction industry in the

DOI: 10.1201/9780429243226-9

UAE utilizes almost half of the resources and raw materials available (Govindan et al., 2016). In addition, the industry alone accounts for 9% of the water consumption, which being poorly acknowledged as the sustainable resource in the construction industry.

Therefore, to meet the water demand of these sectors, UAE harbours several conventional and nonconventional sources of water, deployed within its water supply management system. The conventional natural sources include surface water and groundwater whereas the non-conventional sources represent desalinated and treated waste water that use fossil fuels as the energy inputs. However, strikingly low availability of natural water resources has encouraged UAE to meet its requirements through desalination plants that account for 22% of the water produced in the UAE (Murad et al., 2007). Although demands are satisfied with the current water resources, UAE is set to face challenges in the future owing to the depletion of natural water sources, exhaustion of fossil fuels, population growth, increasing urbanization and the climatic consequences of global warming (Shahin and Salem, 2015). This makes forecasting extremely indispensable in determining the demands of UAE in the face of these unprecedented factors, to secure water demand in the future. As forecasting technique helps in developing a pre-notion about the water demand that will help in achieving a breakeven between the water supply and water demand. Thus, the intent of this study is to identify the factors that affect water demand in the UAE and to propose an accurate water demand forecasting model as a form of criterion validity *which compares a variable against the other variable as a direct measure of the concept* that can integrate all important factors relevant to the water management strategies in the UAE (Fink, 2010). Furthermore, it is vital to evaluate the quality of the forecasting model, termed as predictive validity by adopting multiple methods such R^2 and Mean Square Error (MSE) to validate the water demand forecasting model for the UAE.

6.2 Water demand forecasting in the UAE

Demand forecasting in the water supply networks, sets the groundwork for the functioning of the overall system. It is therefore critical to predict the consumer water requirements for the supply network in order to optimize the operation of the water management system. Moreover, Murad et al. (2007) assert that forecasting unveils the consumption patterns, which helps in extrapolating spatial or temporal patterns of water consumption, thereby allowing the appropriate utilization of water resources.

Literature reports on the number of studies that looked into water demand forecasting in the UAE such as Murad et al. (2007), Mohamed and Al-Mualla (2010) and Younis (2016). Two of these studies focused on the city of Umm-Al Quwain (Murad et al., (2007); Mohamed and Al-Mualla (2010)) and another on Al Ain city (Younis (2016)). The studies considered a number of factors that are likely to impact on the future of water demand which include temperature, rainfall, humidity, Consumer Price Index (CPI), GDP and population. All of these studies used different modelling techniques with relevant validity and reliability approaches to ensure accurate water demand predictability.

This study, however, focuses on United Arab Emirates with the plants under investigation include the Department of Energy Abu Dhabi, Dubai Electricity and Water Authority (DEWA), Sharjah Electricity and Water Authority (SEWA) and the Federal Electricity and Water Authority (FEWA) using Long Short-Term Memory (LSTM) model as indicated by Van et al. (2015). Long Short-Term Memory (LSTM) model, a form of Recurrent Neural Networks (RNNs), is an advanced model which is able to learn and memorize the data as it arrives as well as being able to predict texts and draw relationships between variables with substantial accuracy, while considering temperature, rainfall, humidity, Consumer Price Index (CPI), GDP and population factors.

6.2.1 LSTM forecasting model – predictive validity

The LSTM is an architecture or model that uses an artificial recurrent neural network which is used to facilitate deep structural learning (Paul et al., 2019). A typical LSTM unit is made up of an activation function (output gate), input gate and a cell (neuron). The role of a cell is to keep the memory of values over subjective time breaks (Younis, 2016). The input gate and the output gate, on the other hand, regulate how data flows in and out of the cell. All the neurons, input and output gates make up the overall number of hidden layers and the overall network structure. The core of an LSTM model is its cell state, which acts as the conveyor belt to flow the information along the entire chain (Park and Kim, 2019). The model operates in four steps such as: Firstly, the activation function of the LSTM structure decides the type of information to let out of the cell state (neuron) as shown in equation (6.1a). Secondly, the input gate layer (activation function) determines the type of information to let into the cell state and updates the value as in equation (6.1b). The tan h layer then generates a vector of the novel information to add to the cell state. Thirdly, the LSTM model combines the two added information. It updates the content of the cell accordingly to create a new cell state. An operation is then generated to scale the state value and offer potential solutions in equation (6.1c).

Finally, the model determines what to give as an output or solution based on the filtered version of what it has in the cell state. The output gate layer (with its activation function) does complete this task in an organized fashion such that it only produces what it intended to output as shown in equation (6.1d)

$$f_t = \sigma\left(W_f \cdot [h_{t-1}, x_t] + b_f\right) \qquad (6.1a)$$

$$f_t = \sigma\left(W_f \cdot [h_{t-1}, x_t] + b_f\right) \qquad (6.1b)$$

$$C_t = f_t * C_{t-1} + i_t * C_t' \qquad (6.1c)$$

$$o_t = \sigma\left(W_o\left[h_{t-1}, \, x_t\right] + b_o\right), \, h_t = o_t * \tanh(C_t) \tag{6.1d}$$

Source: (Park and Kim, 2019)

whereas in equations (6.1a)–(6.d) x_t acts as an input vector, h_t: hidden layer vector, o_t: output vector, W, b, C: parameter matrices and vector and σ: Activation functions.

Moreover, forecasting model adopts a form of criterion validity that compares a variable or factor against the other variable which is supposed to be a direct measure of the concept (Fink, 2010). The criterion validity is further divided into two categories: *predictive validity* and *concurrent validity*, where the predictive validity is an essential measure of the quality of the forecasting model (Musselwhite and Wesolowski, 2018). Conducting predictive validation or predictive validity refers to the extent to which a variable measure forecasts future performance of the forecasting model and allows to determine essential information and quality of the model. Predictive validity is often considered in conjunction with the concurrent validity that defines the association between the measure and the criteria (Ivanescu, et al., 2016).

Therefore, this study adopts forecasting model and predictive validity to propose an accurate water demand forecasting technique based on the LSTM that incorporates all significant factors to predict the future water demands of the UAE. However, different method can be adopted to evaluate predictive validity of forecasting model such as R^2 known as the coefficient of determination that refers to the total amount of variance in the outcome measurements explained by the forecasting and predictive model. The value of R^2 ranges from 0 to 1, where value closer to 1 indicates better prediction. Another used method is the Mean squared error which refers to the average of the squared differences between the observed values and the predicted, where a smaller MSE value indicates that the difference between the predicted values and the observed values is closer thus indicating a better prediction (Ivanescu, et al., 2016). Thus, for the purpose of this study both the measures of predictive validity have been adopted. The research of this chapter will therefore discuss the methodological steps adopted to conduct this study, with a particular focus on validity techniques used for the model.

6.3 Methodological steps

To address the challenges in the future owing to the depletion of natural water sources, exhaustion of fossil fuels, population growth, increasing urbanization and the climatic consequences of global warming in UAE. The study adopts a quantitative approach to propose an accurate water demand forecasting technique that incorporates all significant factors to predict the future water demands of the UAE. The methodological approach was based on three main steps.

Stage I – The study obtains its primary data for water consumption from the Federal Competitiveness and Statistics Authority over the period of ten years ranging from 2007 to 2017.

Stage II – Predictive Validity is adopted to determine the relationship between the independent and dependent variable and to evaluate the quality of the forecasting model in the following steps:

- Regression analysis (Hypothesis Testing) is adopted to identify the significant factors that impact the water demand forecasting in UAE. A set of six hypotheses is drawn to determine the significant impact of temperature, rainfall, humidity, CPI, GDP and population growth to on water demand forecasting in the UAE.
- Long Short-Term Model (LSTM) is proposed and validated by using Mean Square Error (MSE) to develop a forecasting model to that predicts the future water demand in the UAE.

The rest of this chapter will give an overview about these steps, with a particular focus on validity and reliability of water demand forecasting model.

6.4 Data collection and analysis

This section describes the steps followed for data collection and analysis in order to predict the water forecasting demand in the UAE.

6.4.1 Stage I – Water consumption data in the UAE

The data of water consumption in the UAE was obtained from the Federal Competitiveness and Statistics Authority database over the period of ten years ranging from 2007 to 2017 (Fcsa.gov.ae, 2020). The water consumption database considered several variables or factors, such as the installed capacity of desalinated water plants by entity, quantity of produced water by producer, percentage distribution of produced water quantity by source of water, quantity of used water, desalinated water quantity provided to authorities by provider entity and water tariff by daily consumption slab rate and sector. It also reflects the association of water production and consumption across different sectors. The data was adopted by this study to deduce a causal relationship between the independent variables and dependent variable (water demand) and to develop an LSTM model by training and testing the model to forecast the future water demand in the UAE.

6.4.2 Stage II – Predictive validity

This study adopts forecasting model and predictive validity to propose an accurate water demand forecasting technique based on the LSTM that incorporates all significant factors to predict the future water demands of the UAE by adopting R^2

and MSE as a measure of the quality of forecasting model. Predictive validity was conducted into two steps such as: Step I – R^2 was adopted to deduce a causal relationship between the independent variables (rainfall, humidity, temperature, CPI, GDP and population growth) and dependent variable (water demand) by adopting regression analysis (hypothesis testing), followed by Step II – implementing the LSTM model and calculating the RMSE of the model to determine the predictive validity of the model.

6.4.3 Step I – Regression analysis (hypothesis testing)

The regression analysis is adopted to deduce a causal relationship between the independent variables (rainfall, humidity, temperature, CPI, GDP and population growth) and dependent variable (water demand) by adopting the hypothesis testing. The significance of this step is to remove the redundant factors as such features could lead to an increase in errors in the model that is going to be trained/fitted. Therefore, based on the findings of the literature, the study draws six hypotheses to determine the significance of factors such as temperature, rainfall and humidity, CPI, GDP and population growth on the water demand in the UAE as shown below.

> **Hypothesis 1:** The change in temperature has a high impact on water demand in UAE.
> **Hypothesis 2:** The change in rainfall amount has a high impact on water demand in UAE.
> **Hypothesis 3:** The change in humidity has a high impact on water demand in UAE.
> **Hypothesis 4:** The change in Gross Domestic Product has a high impact on water demand in UAE.
> **Hypothesis 5:** The change in Consumer Price Index has a high impact on water demand in UAE.
> **Hypothesis 6:** The change in Population Growth has a high impact on water demand in UAE.

The findings suggested that CPI, GDP and population growth are the significant factors that impact water demand forecasting in UAE due to their R^2 higher than 0.5 (closer to 1) with p value less than 0.05. However, for the factors such as temperature, humidity and rainfall the p-value was greater than 0.05 and R^2 was found to be less than 0.5 (closer to 0) which indicates little or no effect on the water demand in the UAE. Therefore, they were removed from the data.

6.4.4 Step II – LSTM model

The stage aims to propose a forecasting model based on the LSTM taking into consideration the factors that have an impact on the water demand forecasting in the UAE.

6.4.4.1 Training

The stage aids to train the LSTM model based on the parameters considered as significant predictors of water demand in the UAE. The training stage includes the following steps importing, normalizing, dividing the data to training and validation data, defining the model parameter and launching the fitting command. The first step is importing the restructured data where the columns like the instance number and the year, which is irrelevant for fitting the model will be eliminated. Secondly, the best practices for training neural networks are to normalize the data before fitting and training. As normalizing the data speeds up learning and aids faster convergence of error. The study will adopt tan h activation function which makes it more preferable to use normalized data because this will speed up the training (Van et al., 2015). After normalization, the data was divided into training and validation/testing data. Training data was used exclusively to train the model, while the testing data was used to compute the RMSE and thus evaluate the reliability of the model. A 9/11 percentage split was used. Since the database consists of 11 years (2007–2017), the years from 2007 to 2015 will be used for training/fitting and 2016–2017 for testing.

The model parameters were set as following: as *random number was set to 1, number of features as 4, number of hidden layers between 1 and 200, max training epochs to 250 and activation function as tan h. The number of hidden layers varied between 1 and 200 as trial and error to find the best network architecture.* Finally, after getting the data ready and defining the model parameters, the "trainNetwork" function of Matlab was used to start the internal fitting/training algorithm to obtain file, which defines the finally fitted LSTM model as shown in Figure 6.1. Network architectures between 1 and 200 hidden layers were fitted and the best architecture was chosen based on the RMSE value.

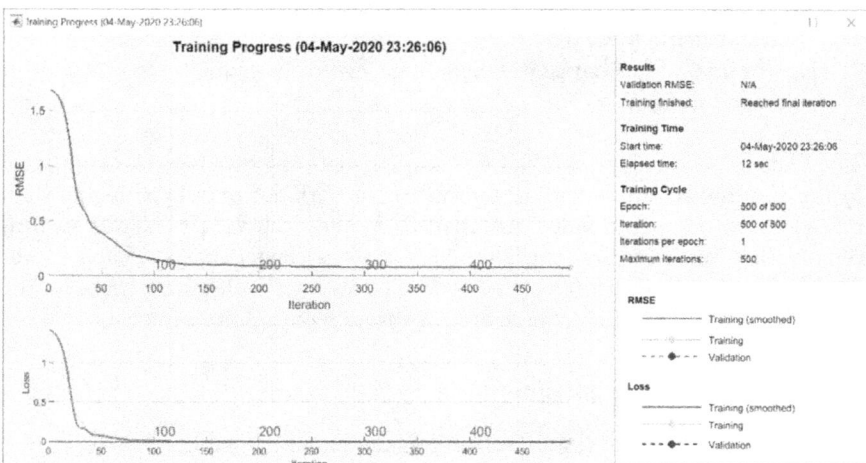

Figure 6.1 Training curves for RMSE and loss

6.4.4.2 Testing (reliability and validity)

The LSTM is a model that uses an artificial recurrent neural network used to facilitate deep structural learning. However, different network architectures (number of hidden layers) lead to different models with varying levels of reliability. This stage, therefore, discusses the process that was used to test different network architectures before finally reaching a conclusion about which architecture is the most superior/reliable.

To evaluate the predictive validity of each model, the RMSE is calculated for both the model as a whole as well as the water demand. The RMSE of the model as whole is calculated by evaluating the error between predicted and actual for all four features (since all features are output due to the nature of the LSTM operating as a recurrent) as shown in equation (6.2) (Gerardnico.com, 2020).

$$\left(\text{RMSE} \mid \text{RMSD}\right) = \sqrt{\frac{\Sigma_{i=1}^{N} \left(Y_i - Y^{\wedge}_i\right)^2}{N}} \tag{6.2}$$

RMSE Equation, Source: (Gerardnico.com, 2020).

The RMSE of the water demand is specific for the focus feature (water demand). This is calculated by measuring the RMSE between predicted and actual for water demand solely. To consider both RMSE, a weighted score for both RMSE was calculated. Each error component was assigned a weight of 0.5 (equal importance), and the sum product was obtained as a weighted score. For all the models, the weighted score RMSE was calculated and sorted starting with the model with the least error as shown in Figure 6.2.

As seen in Figure 6.4, the architecture with 18 hidden layers has the least weighted RMSE score of approximately 0.12, thus considered the most superior/reliable. A graph representing the water demand, model and weighted RMSE was generated as shown in Figure 6.3.

The first line in Figure 6.3(a) represents the water demand RMSE, the last line represents model RMSE and the middle line represents the weighted RMSE. As seen above, models between 4 and 30 hidden layers have the minimum errors in terms of water demand as well as the weighted error. For better visualization, emphasis is put on architectures between 1 and 30 hidden layers as shown below in Figure 6.3(b). As a side note, RMSE values below 0.2 are considered highly reliable, while values below 0.5 are considered acceptable (ResearchGate, 2020). All architectures between 4 and 30 hidden layers are considered excellent because their weighted RMSE score is below 0.2. Nevertheless, the network architecture with 18 hidden layers proves to be the superior/reliable architecture. Furthermore, Figure 6.3(c) portrays the training process for the most superior architecture (18 hidden layers). The loss function diverged to almost zero and the RMSE was almost below 0.1. This shows that the training error was very low, and that the training reliability is high. Taking into consideration that training RMSE was below 0.1, model RMSE was 0.23, weighted RMSE 0.12 and most importantly the water demand RMSE 0.015, it

	A	B	C	D	E
1	N hidden layers	Water demand RMSE	Model RMSE	Weighted score	
2	18	0.015094428	0.23135442	0.123224424	
3	14	0.006204872	0.24492233	0.125563601	
4	5	0.001441093	0.2698265	0.135633796	
5	17	0.050370008	0.22497737	0.137673689	
6	7	0.010679631	0.26781377	0.139246701	
7	19	0.035547644	0.24480891	0.140178277	
8	26	0.078197271	0.20332852	0.140762896	
9	12	0.038432747	0.24893227	0.143682509	
10	8	0.01526299	0.28415877	0.14971088	
11	25	0.085007459	0.21706831	0.151037885	
12	23	0.085495144	0.21740362	0.151449382	
13	6	0.070830256	0.23392841	0.152379333	
14	15	0.074789792	0.23061863	0.152704211	
15	16	0.055522949	0.2523925	0.153957725	
16	20	0.10301235	0.21417356	0.158592955	
17	21	0.10682037	0.21324751	0.16003394	
18	13	0.085105926	0.23531592	0.160210923	
19	28	0.11940554	0.20650202	0.16295378	
20	9	0.09126392	0.23547643	0.163370175	

Figure 6.2 Evaluating best architecture

Source: Authors

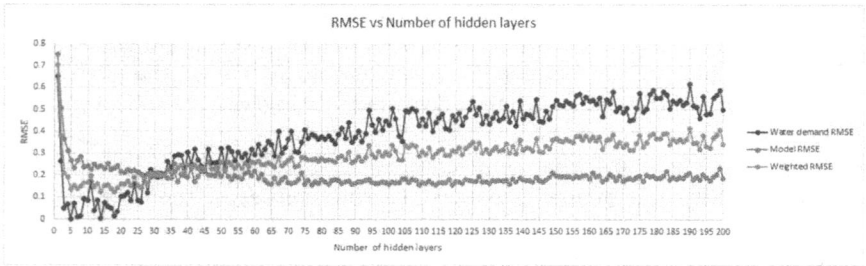

Figure 6.3(a) RMSE vs number of hidden layers (full range)

Source: Authors

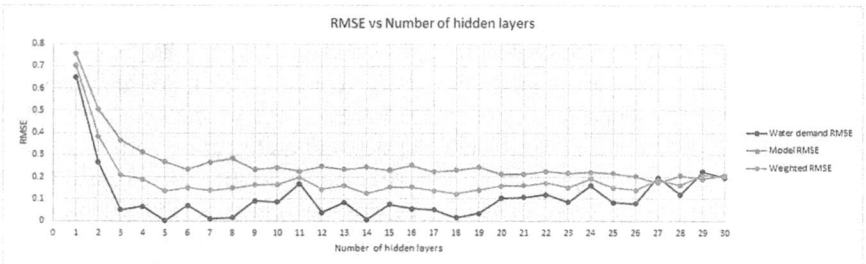

Figure 6.3(b) RMSE vs number of hidden layers (1–30)

Source: Authors

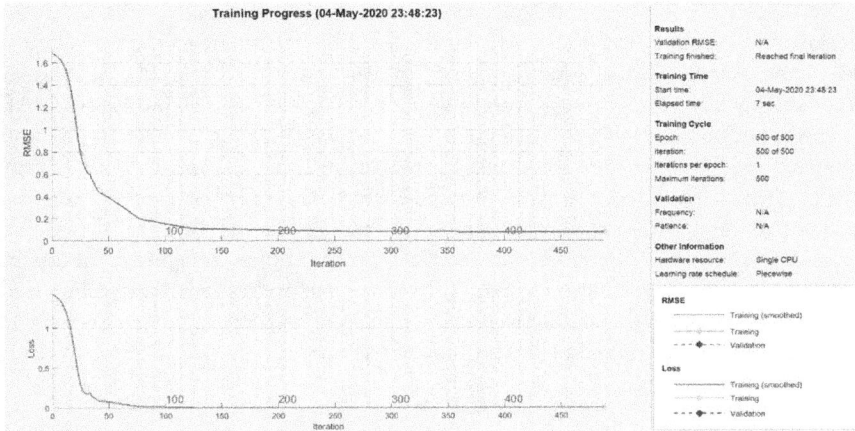

Figure 6.3(c) Training curves for most superior architecture

Source: Authors

can be concluded that the 18 hidden layers network architecture is highly superior and in fact the most superior from all other architectures. Therefore, the chosen LSTM model can be confidently considered to be reliable and accurate enough to be used later for forecasting the water consumption for the next ten years.

6.4.5 Generated output

The chosen architecture (18 hidden layers) was trained and the chosen LSTM model was run for the next ten years from 2018 to 2027. According to the predictions done using the chosen LSTM deep learning model, the water consumption in the UAE for the next ten years is expected to decrease. As shown in Figure 6.4, water consumption is decreasing from 1821.6 million m^3 in 2018 to 1809.9 million m^3 in 2027.

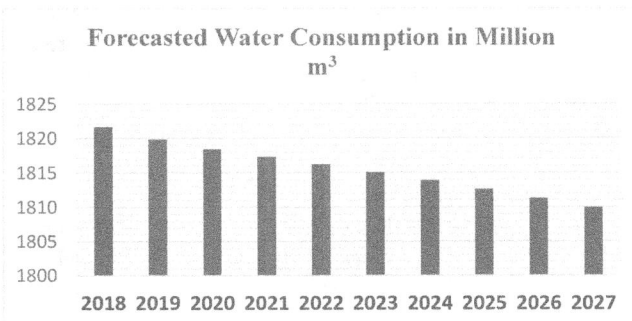

Figure 6.4 Forecasted water consumption for the next ten years

Source: Authors

The decrease in the forecasted water consumption especially after 2016 can be justified by the great efforts put by the UAE government to raise awareness regarding sustainability, establishments of water saving programs as well as imposing regulations for illegal water use. According to Bardsley, (2018), the UAE government had imposed a new regulation regarding illegal groundwater use and required all farms to install metres. Moreover, in 2017 the ministry of Energy and Industry launched the UAE water security strategy that targets reducing the yearly water consumption per capita in the UAE to the half by 2036 (UAE Government, 2020). Therefore, the decreasing trend shown in Figure 6.4 is attributed to factors that are more influential to water demand compared to CPI, GDP and population growth. Be that as it may, the model predicted the decrease in the water demand which proves the accuracy of the model.

6.5 Conclusion

The United Arab Emirates has been suffering from water stress due to the lack of surface water, lack of rainfall and the excessive pumping of groundwater. Thus, due to this it has shifted to desalination projects to help satisfy future demands which are costly in terms of maintenance and operation. Therefore, the need to have an accurate water demand forecasting technique was deemed essential to optimize the water management system and thereby tackling the research gap. Therefore, this study adopts forecasting model and predictive validity to propose an accurate water demand forecasting technique based on the LSTM that incorporates all significant factors such as mean temperature, mean rainfall, relative humidity, CPI, GDP and population growth identified from the literature to predict the future water demands of the UAE. However, to identify the significant predictors of water demand forecasting, six hypotheses were formed, that aid in identify CPI, GDP and population growth as the significant factors affecting the water demand in UAE. The data from an open source provided from the Federal Competitiveness and Statistics Authority over the period of 2007–2017 was used to train the LSTM model using MATLAB software. The LSTM model was trained with 9/11 percentage split of the data (2007–2015) and tested with 2/11 percentage split of the data (2016–20017). The results of the LSTM model and predictive validity indicated the model is highly reliable and valid and showed a decreasing trend from 1821 million m^3 in 2018 to 1809.9 million m^3 in 2027.

References

Ait-Kadi, M., 2016. Water for development and development for water: Realizing the sustainable development goals (SDGs) vision. *Aquat Procedia*, 6, pp. 106–110. DOI: https://doi.org/10.1016/j.aqpro.2016.06.013

Alsharhan A.S., Rizk Z.E. (2020) Water resources and water demands in the UAE. In: *Water Resources and Integrated Management of the United Arab Emirates. World Water Resources* (vol. 3). Springer, Cham. DOI: https://doi.org/10.1007/978-3-030-31684-6_27

Al-Qawasmi, J., Asif, M., El Fattah, A.A. and Babsail, M.O., 2019. Water efficiency and management in sustainable building rating systems: Examining variation in criteria usage. *Sustainability, 11*(8), p. 2416. DOI: https://doi.org/10.3390/su11082416

Bardsley, D., 2018. "UAE making major efforts to overcome water conservation's 'many challenges'," https://www.thenational.ae, 02-Feb-2018. [Online]. Available: https://www.thenational.ae/uae-making-major-efforts-to-overcome-water-conservations-many-challenges-1.692196. [Accessed: 05-May-2020].

Boretti, A. and Rosa, L., 2019. Reassessing the projections of the world water development report. *NPJ Clean Water, 2*(1), pp.1–6. DOI: https://doi.org/10.1038/s41545-019-0039-9

Fcsa.gov.ae, 2020. "Home". [Online]. Available: https://fcsa.gov.ae/en-us. [Accessed: 07-Apr-2020].

Fink, A. 2010. Survey research methods, In *International Encyclopedia of Education* (pp. 152–160), P. Peterson, E. Baker, B. McGaw (Eds), (Third Edition), Elsevier, ISBN 9780080448947. DOI: https://doi.org/10.1016/B978-0-08-044894-7.00296-7

Gerardnico.com, 2020, "Data mining - Root mean squared (Error|Deviation) (RMSE|RMSD) [Gerardnico - The Data Blog]". Available: https://gerardnico.com/data_mining/rmse. [Accessed: 04-May-2020].

Gosling, S.N. and Arnell, N.W., 2016. A global assessment of the impact of climate change on water scarcity. *Climatic Change, 134*(3), pp. 371–385. DOI: https://doi.org/10.1007/s10584-013-0853-x

Govindan, K., Shankar, K.M. and Kannan, D., 2016. Sustainable material selection for construction industry–A hybrid multi criteria decision making approach. *Renewable and Sustainable Energy Reviews, 55*, pp. 1274–1288. DOI: https://doi.org/10.1016/j.rser.2015.07.100

Ivanescu, A.E., Li, P., George, B., Brown, A.W., Keith, S.W., Raju, D. and Allison, D.B., 2016. The importance of prediction model validation and assessment in obesity and nutrition research. *International Journal of Obesity, 40*(6), pp. 887–894.

Mohamed, M.M. and Al-Mualla, A.A., 2010. Water demand forecasting in Umm Al-Quwain (UAE) using the IWR-MAIN specify forecasting model. *Water Resources Management, 24*(14), pp.4093–4120. DOI: https://doi.org/10.1007/s11269-010-9649-1

Murad, A.A., Al Nuaimi, H. and Al Hammadi, M., 2007. Comprehensive assessment of water resources in the United Arab Emirates (UAE). *Water Resources Management, 21*(9), pp.1449–1463. DOI: https://doi.org/10.1007/s11269-006-9093-4

Musselwhite, D.J. and Wesolowski, B.C., 2018. *The SAGE encyclopedia of educational research, measurement, and evaluation*. Thousand Oaks, California: SAGE Publications, Inc. ©2018.

Paul, I.J.L., Sasirekha, S., Vishnu, D.R. and Surya, K., 2019, April. Recognition of handwritten text using long short term memory (LSTM) recurrent neural network (RNN). In *AIP Conference Proceedings* (Vol. 2095, No. 1, p. 030011). AIP Publishing LLC. DOI: https://doi.org/10.1063/1.5097522

Park, S. and Kim, Y., 2019, June. A method for sharing cell state for LSTM-based language model. In *International Conference on Intelligence Science* (pp. 81–94). Springer, Cham. DOI: https://doi.org/10.1007/978-3-030-25213-7_6

ResearchGate, 2020. "What's the acceptable value of Root Mean Square Error (RMSE), Sum of Squares due to error (SSE) and Adjusted R-square?", [Online]. Available: https://www.researchgate.net/post/Whats_the_acceptable_value_of_Root_Mean_Square_Error_RMSE_Sum_of_Squares_due_to_error_SSE_and_Adjusted_R-square. [Accessed: 07-Apr-2020].

Shahin, S.M. and Salem, M.A., 2015. The challenges of water scarcity and the future of food security in the United Arab Emirates (UAE). *Nat Resour Conserv*, 3(1), pp.1–6. DOI: https://doi.org/10.13189/nrc.2015.030101

UAE Government, 2020. "The UAE Water Security Strategy 2036 – The Official Portal of the UAE Government". [Online]. Available: https://u.ae/en/about-the-uae/strategies-initiatives-and-awards/federal-governments-strategies-and-plans/the-uae-water-security-strategy-2036. [Accessed: 05-May-2020].

Van, L.P., De Praeter, J., Van Wallendael, G., De Cock, J. and Van de Walle, R., 2015, January. Machine learning for arbitrary downsizing of pre-encoded video in HEVC. In *2015 IEEE International Conference on Consumer Electronics (ICCE)* (pp. 406–407). IEEE. DOI: https://doi.org/10.1109/ICCE.2015.7066464

Younis, H.I., 2016. Water Demand Forecasting in Al-Ain City, United Arab Emirates. Available at: https://scholarworks.uaeu.ac.ae/all_theses/188

7 Exploring and confirming project owners' motivations for green building project delivery using construct validity test

Ayokunle Olanipekun

7.1 Introduction

Project owners are important stakeholders that make the decision for the delivery of green building projects (Zhang et al., 2019; Yates, 2014) and, when motivated, promote the delivery of green building projects to a successful end (Olanipekun, 2016). Meanwhile, the specific factors that indicate project owners' motivation for green building project delivery have not yet been scientifically specified to ensure practical application. From the literature, the existing factors that motivate higher education academics (Li et al., 2013), contractors (Qi et al., 2010), and other building stakeholders such as architects and engineers (Ahn et al., 2013; Cole, 2011; Feige et al., 2011) have been investigated and do not represent the motivation of project owners to engage in the delivery of green building projects in practice. Rather, project owners remain burdened by the green building delivery barriers, such as economic infeasibility, limited awareness and inflexible regulations (Gan et al., 2015), without appropriate motivations to overcome them in practice.

The aim of this study is to establish project owners' motivation for the delivery of green building projects. There are many variables of green building motivation in the literature (e.g. Darko et al., 2017). The objectives of scientifically exploring and confirming the specific ones that correspond to project owners' motivation will lead to achieving the aim of this study (Watkins, 2018). Methodologically, the green building motivation variables will be explored and confirmed using exploratory factor analysis (EFA) and confirmatory factor analysis (CFA), respectively, and subsequently, their convergent and discriminant validity will be determined to ensure that they represent project owners' motivation for the delivery of green building projects in practice and theory. In terms of the theoretical significance, the variables derived from green building literature are established as factors of project owners' motivation for using a methodological process that can, in future, be used in a similar investigation in the field of study. Practically, project owners can use the established factors to evaluate their levels of motivation for green building project delivery, while governments can use them to appropriate the right motivations for project owners to increase green building project delivery.

DOI: 10.1201/9780429243226-10

7.2 Literature review

7.2.1 Motivation for green building project delivery

Generally, the concept of motivation explains human behaviour and actions (Weiner, 1996). Regarding project delivery, motivation is defined as the needed means for affecting changes from conventional to green building project delivery (Zhang et al., 2019; Feige et al., 2011). Motivation is necessary to overcome the enormity of green building barriers, such as averseness due to culture and high investment costs, and engage in green building project delivery (Liu et al., 2012). Therefore, project owners need to be properly and adequately motivated to engage in green building project delivery.

The variables of green building motivation stem from awareness of green building benefits, which increases the adoption of green building among building professionals (Häkkinen & Belloni, 2011). Environmental altruism is a variable of green building motivation which manifests in the form of concern for the environment and its protection among contractors in China (Qi, Shen, Zeng, and Jorge, 2010). The carrot and stick approach of the government is another variable; the latter manifests in the form of incentives by the government to attract beneficiary stakeholders to construct green building projects (Liu, 2012). The former manifests as regulations and regulatory pressures by the government to enforce compliance with green building practices (Gou et al., 2013). The functional benefits of green building are also considered as motivations for the delivery of green building projects in practice. These are the functions of green building in-use, such as energy conservation, improved indoor environmental quality and waste conservation (Ahn et al., 2013). Also, a study by Windapo (2014) reveals that the presence of a recognized rating system is a major attraction for developers to construct green building projects in South Africa. Such developers grow their reputation by acquiring sustainability ratings in their projects.

Also, Green building motivations can be divided into different types of projects. According to Sauer and Siddiqi (2009), developers are motivated by density bonus to have more space for residential green building construction and increase profits in the process. Potential home buyers less than twenty-five years old are motivated by financial incentives to procure residential green building projects due to their low-income levels (Ghodrati et al., 2012). For office green building projects, Boyle and McGuirk (2012) reveal that it is a palette of motivation factors, particularly ethical commitments and social responsibility, embedded together. With regards to higher education green building projects, they are motivated by owners' desire for enhanced reputation in the society (Li et al., 2013), much like the developers mentioned earlier. In contrast, the delivery of green building projects for financial gain and environmental protection is of less concern for higher education building owners (Li et al., 2013). Green building motivations also have geographical implications, such as the contextual factors and social climate in the geographical locations of green building projects

Table 7.1 Variables of green building motivation

Descriptor	Brief descriptions
GM1	The improvement in the quality of life in green building projects
GM2	The altruistic or personal moral norms and values that are pro-environmental and provoking green building intent
GM3	The enhancement of reputation for green building ownership
GM4	The persuasive influence of green advocacy champions or leaders
GM5	The market appeal of green building projects
GM6	The financial incentives or monetary gains provided by the government
GM7	The non-financial incentives or non-monetary gains provided by the government

(Mason et al., 2011). According to these authors, contextual factors, such as leadership commitment to sustainability issues through signed agreements, propel residents to engage in green building project delivery. Additionally, smaller cities adjacent to cities with a high rate of green building construction are likely to engage in green building construction as well. Furthermore, shared community culture, especially linguistic affiliation, and high-income levels are strong are strong motivation factors for green building projects (Swidler, 2011). From the literature perspective, the variables of green building motivations (GM1 to GM7) are listed in Table 7.1. The variables will be verified by employing a quantitative research methodology to obtain the understanding of green building motivation from the perspectives of project owners.

7.3 Methodology

Achieving the aim of this study is embedded in the positivism research paradigm which limits the creation of knowledge to what can be observed and measured (Krauss, 2005) and the quantitative methodology that permits the use of statistical methods to analyse data obtained about measurables (Abawi, 2008). The variables of green building motivation (see Table 7.1) are operationalizable measures, thereby aligning with the positivism research paradigm and quantitative methodology. To implement the research paradigm and methodology, green building project owners have been identified as respondents to provide responses to GM1–GM7, based on their experiences of green building project delivery in the Australian construction industry. Meanwhile, the Green Star Accredited Professionals (GSAP) normally represents the project owners' interests in green building project delivery in this country (Dimovski & O'Neill, 2015). Also, because they are more accessible to provide responses on behalf of the project owners, 262 GSAP are selected as proxy in similar manner as the proxy technique that has been successfully employed in past built environment studies (e. g. Ng & Tang, 2010). A well-structured questionnaire was administered online to the GSAP (on behalf of project owners) to provide data about how they are influenced by GM1–GM7 to deliver green building projects in practice. GM1–GM7 were ranked on a five-point Likert scale that ranged from 1 as "very low influence" to 5 as "very high influence", thereby making it easier for the GSAP to decide their viewpoints easily (Chew, 2013).

To execute the objectives of exploring and confirming the variables of green building motivation (GM1–GM7), the EFA and CFA methods were employed for the analysis of 150 responses (or data) obtained from the GSAP. Regarding the EFA, the principal axis factoring (PAF) was used for data extraction while the interpretation cut-off was pegged at 0.4 for factor loadings (De Winter & Dodou, 2012). Also, the oblique method of factor rotation was used to allow for factor correlation (Hooper, 2012). Subsequently, the CFA was used to confirm the factors derived from the EFA. According to Albright (2006), an important step in the CFA is model fitting to evaluate the match between the hypothesized model and observed data using fit indices. The commonly used fit indices include the normed fit index (NFI), goodness-of-fit index (GFI), relative fit index (RFI), adjusted goodness-of-fit index (AGFI) and comparative fit index (CFI), and a model is fit when these indexes are >0.9 (Byrne, 2010). Additionally, the convergence (or convergent validity) and discrimination (or discriminant validity) of the factors is confirmed following the CFA. Convergent validity is acceptable when the standardized regression weights of specific items are significant and higher than 0.5 (Xiong et al., 2015). Discriminant validity is affirmed when the average variance extracted (AVE) of a factor is higher than its squared correlations with other factors (Hon et al., 2012).

7.4 Results and analysis

The EFA with PAF extraction method reveals a Kaiser–Meyer–Olkin value of 0.791, which indicates a sufficient sampling adequacy (Williams et al., 2012). Also, it reveals a Bartlett's test of sphericity values ($\chi^2 = 250.464$; df = 21) that is significant at $p < 0.01$, further indicating sufficiently large correlations for EFA (Hooper, 2012). Furthermore, the PAF with oblique rotation reveals the presence of two factors with eigenvalues greater than 1. The factors explain a total variance of 63.23%, with factors 1 and 2 accounting for 47.77% and 15.45%, respectively. As shown in Table 7.2, three variables, namely, GM1, GM2, GM3, & GM4, load under factor 1, while GM5, GM6 & GM7 load under factor 2. Both factors are consistent with internal (INT) and external (EXT) motivations in (Olanipekun et al., 2016) and are labelled as such.

The results show that the CFA model is well fitting with the data given: $\chi^2 = 10.056$; $\chi^2/df = 0.774 < 2$; df = 13; $p = 0.689 > 0.05$; (NFI = 0.968; GFI = 0.959; RFI = 0.948; AGFI = 0.911) > 0.9. Also, the observed variable paths to the latent factors are significant ($p < 0.05$), ranging from 0.758 to 0.938. Regarding the convergent validity, all the standardized regression weights return highly significant values above 0.5, which indicates an acceptable convergent validity of factors (Xiong et al., 2015). Regarding the discriminant validity, the AVE of factors (INT = 0.649) and 2 (EXT = 0.736) are greater than the squared factor correlation between INT and EXT ($R^2 = 0.230$). It indicates dissimilar constructs.

Table 7.2 Pattern and structure matrix for EFA

Item number	Pattern coefficients		Structure coefficients	
	1	*2*	*1*	*2*
Factor 1, internal motivations for delivering green building projects (eigenvalue = 3.344; percentage of variance = 47.777; cumulative percentage = 47.777)				
GM1	**0.801**	0.055	**0.828**	0.443
GM2	**0.818**	0.054	**0.844**	0.450
GM3	**0.788**	−0.110	**0.735**	0.272
GM4	**0.633**	0.077	**0.670**	0.383
Factor 2, external motivations for delivering green building projects (eigenvalue = 1.082; percentage of variance = 15.454; cumulative percentage = 63.231)				
GM5	0.091	**0.722**	0.440	**0.766**
GM6	−0.097	**0.946**	0.360	**0.899**
GM7	0.056	**0.754**	0.420	**0.781**

Note: The major loadings are in bold

7.5 Discussion

Based on the results of the EFA and CFA, GM1–GM7 have been divided into two factors and labelled as internal (INT) and external (EXT) motivations in line Olanipekun et al. (2016). The INT suggests that project owners are driven by volition or personal endorsement towards the delivery green building projects (Olanipekun et al., 2016; Ryan & Deci, 2000). Additionally, INT characterizes the fulfilment of certain psychological needs in terms of values, norms, beliefs or social concerns by the delivery of green building projects (Aliagha et al., 2013; Joachim et al., 2015; Olanipekun, 2016; Olanipekun et al., 2016). In contrast, EXT suggests that project owners are driven towards the delivery of green building projects in order to obtain some separable outcomes such as tangible rewards or benefits (Olanipekun et al., 2016; Vallerand, 2004). Based on the eigenvalues obtained from the EFA, the INT has an eigenvalue of 3.344 and accounting for 47.777% of the variance in the dataset, while the EXT has an eigenvalue of 1.082 and accounting for 15.454% of the variance in the dataset. Therefore, the INT of project owners could be regarded as more important motivation than EXT for the delivery of green building projects.

The discriminant validity test reveals that INT and EXT motivations of project owners for green building project delivery are different from one another. This is consistent with the prevailing theoretical position in the psychology literature that EXT motivation undermines the INT motivation of an individual to perform an action (Deci et al., 1999; Koger et al., 2011). According to Gagné and Deci (2005), providing external and/or tangible incentives; for instance, GM6 diminishes the attempts to perform an action out of volition. In practice, it is often found that homeowners install sustainable features in their homes purposely to receive monetary grants from governments without commitments to embrace sustainable building practices out of their volition (Saka et al., 2021). By implication, this is why many incentivization programmes of governments such as the UK Green Deal have failed (Thorpe, 2016), and not resulted in a significant uptake of green building project delivery among project owners. As a result, combining INT and EXT motivations to ensure that the advantages in one complements the disadvantages in the other for an increased green building project delivery has been suggested in the literature (see Olanipekun et al., 2018).

7.6 Conclusion

This study finds that the variables of green building motivation in the literature can be scientifically explored and confirmed to establish project owners' motivation for green building project delivery. Following data collection from GSAP (on behalf of project owners) and the analysis of data using EFA and CFA, this paper concludes that project owners' motivation for green building project delivery can be divided into internal (INT) and external (EXT) factors of motivation. Also, based on the results of the discriminant validity of both factors of motivation, this paper concludes that the external motivation undermines the internal one in practice, which is consistent with the literature on psychology of motivation. Despite that project owners can be motivated by two different factors of motivation, the implication is that both should be applied distinctively to prevent counter-productivity in green building project delivery. Alternatively, individual project owners can key into their internal motivation for green building project delivery, while governments can provide external motivations, such as monetary grants for project owners, to engage in green building project delivery. Clearly, both internal and external factors of motivation have the merits of increasing green building project delivery. In future, both internal and external motivations can be explored and how they can be combined to increase green building construction among project owners and other stakeholders can be demonstrated.

References

Abawi, K. (2008). Qualitative and quantitative research. World Health Organization/ Geneva Foundation for Medical Education and Research, Geneva. http://www.gfmer. ch/Medical_education_En/Afghanistan_

Ahn, Y. H., Pearce, A. R., Wang, Y., & Wang, G. (2013). Drivers and barriers of sustainable design and construction: The perception of green building experience. *International Journal of Sustainable Building Technology and Urban Development*, 4(1), 35–45.

Albright, J. J. (2006). *Confirmatory Factor Analysis Using AMOS, LISREL, and MPLUS*. USA: The Trustees of Indiana University. Available at: http://www.iub.edu/~statmath/ stat/all/cfa/cfa2008.pdf

Aliagha, G. U., Hashim, M., Sanni, A. O., & Ali, K. N. (2013). Review of green building demand factors for Malaysia. *Journal of Energy Technologies and Policy*, 3(11), 471–478.

Boyle, T., & McGuirk, P. (2012). The decentred firm and the adoption of sustainable office space in Sydney, Australia. *Australian Geographer*, 43(4), 393–410

Byrne, B. M. (2010). *Structural equation modelling with AMOS: basic concepts, applications, and programming* (2nd ed.). Mahwah: Erlbaum.

Cole, R. J. (2011). Motivating stakeholders to deliver environmental change. *Building Research & Information, 39*(5), 431–435.

Chew, K. (2010). Singapore's strategies towards sustainable construction. *The IES Journal Part A: Civil & Structural Engineering, 3*(3), 196–202.

Darko, A., Zhang, C., & Chan, A. P. (2017). Drivers for green building: A review of empirical studies. *Habitat International, 60*, 34–49.

Deci, E. L., Koestner, R., & Ryan, R. M. (1999). A meta-analytic review of experiments examining the effects of extrinsic rewards on intrinsic motivation. *Psychological Bulletin, 125*(6), 627–668.

De Winter, J. C., & Dodou, D. (2012). Factor recovery by principal axis factoring and maximum likelihood factor analysis as a function of factor pattern and sample size. *Journal of Applied Statistics, 39*(4), 695–710.

Dimovski, B., & O'Neill, L. (2015). Sustainability, A-REITs and the global financial crisis. *Pacific Rim Property Research Journal, 21*(1), 51–59.

Feige, A., Wallbaum, H., & Krank, S. (2011). Harnessing stakeholder motivation: Towards a Swiss sustainable building sector. *Building Research & Information, 39*(5), 504–517.

Gagné, M., & Deci, E. L. (2005). Self-determination theory and work motivation. *Journal of Organizational Behavior, 26*(4), 331–362.

Gan, X., Zuo, J., Ye, K., Skitmore, M., & Xiong, B. (2015). Why sustainable construction? Why not? An owner's perspective. *Habitat International, 47*(2015), 61–68. doi: http://dx.doi.org/10.1016/j.habitatint.2015.01.005.

Ghodrati, N., Samari, M., & Shafiei, M. (2012). Investigation on government financial incentives to simulate green homes purchase. *World Applied Sciences Journal, 20*(6), 832–841.

Gou, Z., Lau, S. S. Y., & Prasad, D. (2013). Market readiness and policy implications for green buildings: case study from Hong Kong. *Journal of Green Building, 8*(2), 162–173.

Häkkinen, T., & Belloni, K. (2011). Barriers and drivers for sustainable building. *Building Research & Information, 39*(3), 239–255.

Hon, C. K., Chan, A. P., & Yam, M. C. (2012). Determining safety climate factors in the repair, maintenance, minor alteration, and addition sector of Hong Kong. *Journal of Construction Engineering and Management, 139*(5), 519–528.

Hooper, D. (2012). Exploratory factor analysis. In H. Chen (Ed.), *Approaches to Quantitative Research–Theory and Its Practical Application: A Guide to Dissertation Students*. Cork, Ireland: Oak Tree Press.

Joachim, O. I., Kamarudin, N., Aliagha, G. U., & Ufere, K. J. (2015). Theoretical explanations of environmental motivations and expectations of clients on green building demand and investment. In *IOP Conference Series: Earth and Environmental Science* (vol. 23, pp. 012010). Bristol, IOP Publishing.

Krauss, S. E. (2005). Research paradigms and meaning making: A primer. *The Qualitative Report, 10*(4), 758–770.

Koger, S. M., Leslie, K. E., & Hayes, E. D. (2011). Climate change: Psychological solutions and strategies for change. *Ecopsychology, 3*(4), 227–235.

Li, X., Strezov, V., & Amati, M. (2013). A qualitative study of motivation and influences for academic green building developments in Australian universities. *College Publishing, 8*(3), 166–183.

Liu, Y. (2012). Green building development in China: a policy-oriented research with a case study of Shanghai. *Master Thesis Series in Environmental Studies and Sustainability Science*.

Liu, J. Y., Low, S. P., and He, X. (2012). Green practices in the Chinese building industry: Drivers and impediments. *Journal of Technology Management in China*, 7(1), 50–63.

Mason, S. G., Marker, T., & Mirsky, R. (2011). Primary factors influencing green building in cities in the Pacific Northwest. *Public Works Management & Policy*, 16(2), 157–185.

Ng, S. T., & Tang, Z. (2010). Labour-intensive construction sub-contractors: Their critical success factors. *International Journal of Project Management*, 28(7), 732–740.

Olanipekun, A.O. (2016), The levels of building stakeholders' motivation for adopting green buildings. Being a paper presented at the Joint International Conference organised by the Federal University of Technology, Akure, London South Bank University and De Montfort University, theme: "21st Century Human Habitat: Issues, Sustainability and Development, held in Akure", Nigeria, 2016.

Olanipekun, A. O., Xia, P. B., & Skitmore, M. (2016). Green building incentives: A review. *Renewable and Sustainable Energy Reviews*, 59(June 2016), 1611–1621.

Olanipekun, A. O., Xia, B., Hon, C., & Darko, A. (2018). Effect of motivation and owner commitment on the delivery performance of green building projects. *Journal of Management in Engineering*, 34(1), 04017039.

Qi, G., Shen, L., Zeng, S., & Jorge, O. J. (2010). The drivers for contractors' green innovation: An industry perspective. *Journal of Cleaner Production*, 18(14), 1358–1365.

Ryan, R. M., & Deci, E. L. (2000). Intrinsic and extrinsic motivations: Classic definitions and new directions. *Contemporary Educational Psychology*, 25(1), 54–67.

Saka, N., Olanipekun, A. O., & Omotayo, T. (2021). Reward and compensation incentives for enhancing green building construction. *Environmental and Sustainability Indicators*, 11, 100138.

Sauer, M., & Siddiqi, K. (2009). Incentives for green residential construction. In Construction Research Congress 2009. Building a Sustainable Future, Seattle, WA, USA, 5–7 April 2009; pp. 578–587.

Swidler, S. (2011). Homeownership: Yesterday, today and tomorrow. *Journal of Financial Economic Policy*. 3(1), 5–11.

Thorpe, D. (2016). Why the UK Green Deal failed and why it needs a replacement. *Energy Post*. Retrieved on 24 August, 2021 from Why the UK green deal failed (energypost.eu)

Vallerand, R. J. (2004). Intrinsic and extrinsic motivation in sport. *Encyclopedia of Applied Psychology*, 2(10), 427–435.

Watkins, M. W. (2018). Exploratory factor analysis: A guide to best practice. *Journal of Black Psychology*, 44(3), 219–246.

Weiner, B. (1996). Searching for order in social motivation. *Psychological Inquiry*, 7(3), 199–216.

Williams, B., Onsman, A., & Brown, T. (2012). A Rasch and factor analysis of a paramedic graduate attribute scale. *Evaluation & the Health Professions*, 35(2), 148–168.

Windapo, A. O. (2014). Examination of green building drivers in the South African construction industry: Economics versus ecology. *Sustainability*, 6(9), 6088–6106.

Xiong, B., Skitmore, M., & Xia, B. (2015). A critical review of structural equation modeling applications in construction research. *Automation in Construction*, 49(2015), 59–70.

Yates, J. (2014). Design and construction for sustainable industrial construction. *Journal of Construction Engineering and Management*, 140(4), B4014005.

Zhang, J., Li, H., Olanipekun, A. O., & Bai, L. (2019). A successful delivery process of green buildings: The project owners' view, motivation and commitment. *Renewable Energy*, 138, 651–658.

8 Investigation into underpinning criteria of depression in women by adopting factor analysis as construct validity test

Sara Saboor, Sadawi Alzzatrah and Vian Ahmed

8.1 Background

According to the World Health Organization (WHO), depression is recognized as the most common mental disorder that affects more than 264 million people around the globe. Depression is classified as a form of continuous sadness and loss of interest in daily activities, which results in a lack of sleep and appetite. It is a result of a complex interaction of one's personal (biological, psychological), social and daily aspects of life events (WHO, 2021). Studies have also shown that the chances of depression and anxiety disorder are more common at a younger age and especially in women when compared to men.

Women play a vital role in any society and for the development and success of its built environment. In recent years, studies have shown that women are participating actively as key players in managing and developing their built environment and are likely more capable to anticipate and react to crises (Jabeen, 2014). The studies also show that women make up to 13% of the United Kingdom (UK) construction industry with inspiring women like Zaha Hadid who won the highest architecture award Pritzker Architecture Prize in 2004; Carosline Moser an urban social anthropologist who dedicated 40 years working as a social policy specialist in the built environment and many more. In recent years the contribution of women in society has exponentially increased from the role of women as caretakers only to women as educators, as part of the workforce and as global volunteers (Fairgaze, 2021). According to the UN report, about 72% of the COVID-19 pandemic front line healthcare workers were women, with 1 in 4 women held parliamentary seats around the world, and women participation in the peace process has increased significantly from years 1992 to 2019 as mediators and negotiators (Women, 2019).

Thus, women being vital members of society, it becomes significant to address the issue of mental disorders such as depression that is found to be common in women. Therefore, this study intends to identify the underpinning criteria of women daily life that impacts depression. In addition, the study adopts exploratory and confirmatory factor analysis for the investigation of construct validity to determine the relationship between the identified variables of women's daily life on their mental health, i.e. Depression.

DOI: 10.1201/9780429243226-11

However, due to the huge number of variables and the dynamic nature of the construct it cannot be studied in general but rather as per population or area because it is a diverse problem that one solution cannot fit all. Therefore, this study limited its scope to the United Arab Emirates as the focus of the research. Like other developed countries, studies in the Gulf and Arab region have similar findings, where the studies have shown that depression rates range from 13% to 18% with the female's chance of depression and anxiety disorder doubled as compared to the male in this region (Razzak et al., 2019). A study by McIlvenny et al. (2000) found that fatigue, anxiety and depression in females are more than males in the UAE, while factors such as lack of exercise, obesity and illiteracy play critical roles.

8.2 Literature review

Depression is considered to be the most common form of illness and mental disorder globally that is a result of a complex interaction of one personal, social and biological factors, which is found to be most common in women than men. According to Piccinelli and Wilkinson (2000) the variables that contribute to depression in women remain at large related to different dynamics such as "family environmental factors", "age", and "personal attributes" which act as few of the potential candidates of depression episodes.

As a result, emerging studies have mainly focused on only a few variables at a time and for specific geographical locations. For example, a study conducted by Regestein et al. (2010) focused on evaluating the relationships between sleep habits and depressive symptoms. The study demonstrated that bedtimes specifically later than the critical hour of 2:00 a.m. risked higher Depression Tendency scores. Connections were found between sleep and depression clinical problems such as energy level, self-image, concentration, appetite, etc. Similarly, Lombardo et al. (2014) studied the association between eating disorders and reduction of sleep quality. These findings supported the existence of both a direct and an indirect relationship between insomnia symptoms and eating disorder symptoms.

Likewise, Tong et al. (2014) investigated the prevalence of eating disorders (ED) in female university students in China, using a two-stage design. Another study by Kim and Lee (2013) examined the bi-directional relationship between intimate partner violence (IPV) and depression using prospective data. Data from the Korean Welfare Panel Study (KOWEPS) were used to test whether IPV was associated with an increased overall level of depression and with the rate of change over time in depressive symptoms and whether this model of change in depressive symptoms was associated with subsequent incidences of IPV. Furthermore, the literature also reports on various studies that investigate the impact and relationship of daily, personal and social aspects of a female life on their depression level. In their study, Hussain and Cochrane (2002) suggested that understanding the experience of conflicting cultural expectations, distinctions between psychosocial, spiritual, physical health problems and communication problems (general and culture-specific) were central to the women's experiences of depression.

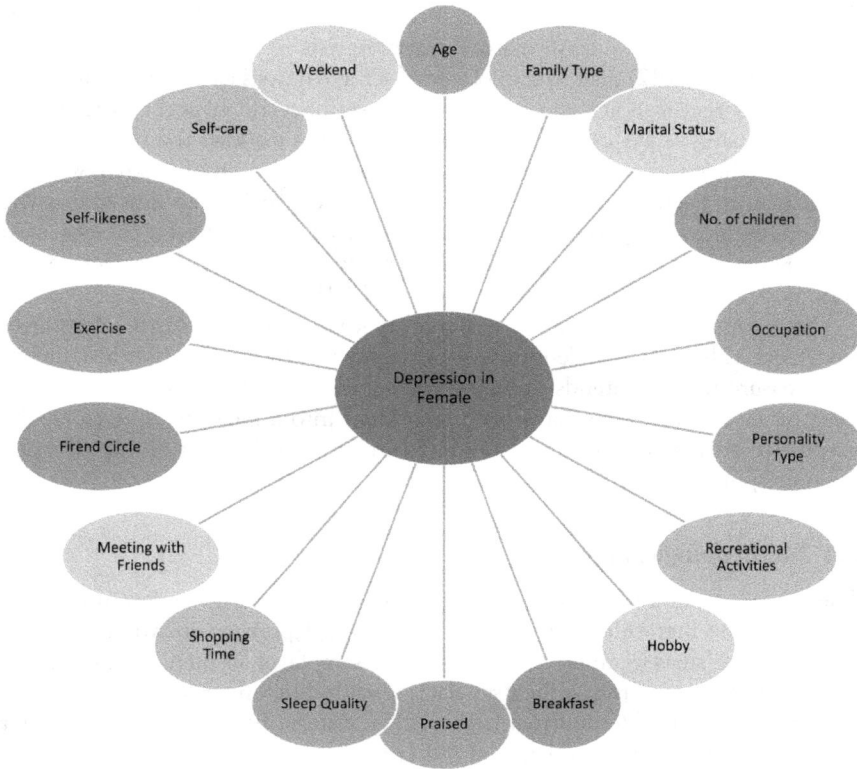

Figure 8.1 Underpinning criteria

As such, the research in hand such as (Hussain and Cochrane, 2002, Sehlo and Bahlas, 2013, Sung et al., 2016) summarizes several variables from the literature that have a significant impact on depression in females as shown in Figure 8.1.

The literature identified 18 underpinning criteria that impact depression in women. However, as the study of depression in women in the United Arab Emirates is relatively new; this study adopts exploratory and confirmatory factor analysis, multivariate statistical techniques to understand these 18 criteria identified as shown in Figure 8.1 in terms of UAE and interprets them into meaningful factors (latent factors).

The next section will therefore elaborate on the techniques used to understand the relationship between the criteria and allow them to be structured (interpreted) into meaningful factors.

8.3 Factor analysis as a reliability measure

Factor analysis is a statistical technique adopted to reduce the dimension or a large number of criteria into fewer factors. The technique allows to investigation the construct validity by aiding the study to understand the relationship between the criteria and factors. It is often categorized into two categories such as exploratory factor analysis and confirmatory factor analysis.

Levant et al. (2012) defines Exploratory Factor Analysis as a preliminary analysis of construct validity when the researchers have no hypothesis regarding the relationship of the criteria and factors. It generally allows the researchers to determine the structure of the factors and examine the internal validity, whereas Confirmatory Factor Analysis is a popular statistical technique that allows the researchers to test hypotheses regarding the relationship between the observed criteria and latent factors. It also allows the study to determine the structure and path model of criteria into factors (Atkinson, 2011).

The study adopts exploratory and confirmatory factor analysis to investigate Construct Validity which determines the degree to which the test or instrument can measure what it intends to be measuring. In addition, it also measures how well the ideas and theory have been translated into a measure. Therefore, the purpose of this study is to understand and investigate the relationship between the underpinning criteria of women's daily life on their depression.

8.4 Methodological approach

The study adopts a mixed-method approach to identify the underpinning criteria from the literature that have an impact on depression in women using a questionnaire survey, targeting a population sample of women in the United Arab Emirates to determine their satisfaction with different criteria of their daily, personal and social life and to whether these criteria have an impact on their depression. Moreover, to understand the criteria and to determine the relationship between them the study adopts exploratory factor analysis with the aid of principal component analysis. This stage intends to reduce the dimension of the dataset and interpret the criteria into meaningful factors. After conducting the exploratory analysis stage, the study applies confirmatory factor analysis to investigate the construct validity by testing the conceptual hypothetical model and structure produced in the exploratory factor analysis stage.

8.5 Data collection and analysis

This section highlights the results of the survey (questionnaire) followed by the exploratory factor analysis and confirmatory factor analysis that was adopted as a measure of construct validity.

8.5.1 Demographic profile

The study adopts a questionnaire survey to determine the impact of criteria of daily, personal and social aspects of a women's life that have an impact on their depression. The survey was distributed to females in a family clinic in Dubai with aid of medical students. The total responses collected were 123 respondents as shown in Table 8.1. The survey aims to determine the satisfaction of females with

Table 8.1 Participants demographics

Characteristics	Range	No. of participants (n = 123)	Depressed	Non-depressed
Age	18–25	64	53	11
	26–34	28	22	6
	35–42	10	7	3
	43–50	21	7	14
Personality type	Introvert	53	39	14
	Extrovert	70	23	47
Marital status	Married	48	28	20
	Single	70	54	16
	Divorced/Widowed	5	4	1
Family type	Nuclear (Husband and children)	65	44	21
	Extended family (with relatives)	58	42	16
Occupation	Working women	45	26	19
	Homemaker	59	47	12
	Freelancer/Other	19	12	8
Children	0	83	62	21
	1	14	9	5
	2	14	10	4
	3	7	4	3
	4	5	1	4
Depression	Yes	84		
	No	37		

different criteria of their life that they believe have an impact on their depression. The survey was divided into three sections.

Firstly, the demographics of the participating females were questioned, secondly, they were asked to self-evaluate if they experience depression on a 5 level Likert scale from (1 – Never to 5 – Always), finally followed by their perception of satisfaction with different criteria of daily, personal and social aspects of their life on 5 levels Likert scale ranging from (1 – Highly dissatisfied to 5 – Highly Satisfied).

Table 8.1 shows that 84 (68%) of the females indicate that they faced little or severe form of depression whereas 37 feel that that they don't experience any form of depression. The table also highlights that females in the age group 18–25, introvert personalities, Single, Homemakers, and females with no children are more prone to experience depression. However, the field of depression in females in the UAE is relatively new. Therefore, to understand the criteria and the relationship between them, it is necessary to adopt an exploratory factor analysis.

8.5.2 Exploratory factor analysis

Exploratory factor analysis is a statistical technique to identify the underlying theoretical structure of variables and their relationships, it also aids in reducing the dimensions of the dataset.

8.5.2.1 Step 1 – Factor analysis – calculating communalities

In order to investigate the construct validity and understanding of the relationship between variables of daily, personal and social aspects of a female's daily life, exploratory factor analysis on the 18 criteria in Figure 8.1 was conducted. The analysis adopts an alpha (significance value) of 0.5 as suggested by Williams, Onsman, and Brown (2010). The alpha or significance value allows determining which criteria will be included or excluded from the analysis, whereby, the communalities are the measure of the amount of variance in each criterion that can be explained by the factors. Hence, the communalities of variables in Table 8.2(a) present the variance explained by the criteria or the information contained by the criteria, while all the variable communalities are above 0.5, indicating they are significant criteria of women's daily life that have an impact on depression.

However, "Age", "Family Type" and "Occupation" were removed from the analysis as the variance was below the threshold value set at 0.5. Thus only 15 criteria were found to be significant such as Personality trait, marital status, no. of children, recreational activity, weekend, self-care, hobby, breakfast, praised, sleep quality, shopping, friend circle, meeting with friends, exercise and self-likeness.

8.5.2.2 Step 2 – Retaining factors – Principal component analysis

The exploratory factor analysis adopts the Principal Component Analysis (PCA) as an extraction method to reduce the dimension and to determine the number of factors to retain as shown in Table 8.2(b), whereas suggested by Habing (2003), Factors having an eigenvalue larger than 1, and those that account for about 70–80% of the total cumulative variance will be retained in the analysis.

Therefore, it can be concluded here that as the first five factors have an eigenvalue greater than 1 and explained 68.617 of the total cumulative % variance will

Table 8.2(a) Communalities

	Extraction
Personality trait	0.623
Marital status	0.734
Children	0.612
Recreational activity	0.549
Weekend	0.734
Self-care	0.656
Hobby	0.552
Breakfast	0.598
Praised	0.636
Sleep quality	0.671
Shopping	0.602
Friend circle	0.77
Meeting with friends	0.65
Exercise	0.542
Self-likeness	0.564

Table 8.2(b) Principal component analysis

Factors	Initial eigenvalues			Extraction sums of squared loadings			Rotation sums of squared loadings		
	Total	% of Variance	Cumulative %	Total	% of Variance	Cumulative %	Total	% of Variance	Cumulative %
1	3.259	21.724	21.724	3.259	21.724	21.724	2.512	16.747	16.747
2	2.662	17.744	39.468	2.662	17.744	39.468	1.97	13.133	29.879
3	1.389	9.259	48.727	1.389	9.259	48.727	1.755	11.698	41.577
4	1.062	7.082	55.809	1.062	7.082	55.809	1.585	10.567	52.145
5	1.021	6.808	62.617	1.021	6.808	62.617	1.571	10.472	62.617
6	0.945	6.3	68.917						
7	0.829	5.526	74.442						
8	0.774	5.16	79.602						
9	0.626	4.17	83.772						
10	0.572	3.815	87.587						
11	0.536	3.572	91.159						
12	0.434	2.892	94.051						
13	0.357	2.382	96.433						
14	0.314	2.094	98.527						
15	0.221	1.473	100						

Table 8.2(c) Rotated matrix

	Component				
	Personal	Self-care	Emotional	Leisure	Health
Marital status	0.833				
Children	0.7				
weekend	0.664				
Meeting with friend	0.538				
exercise		0.64			
self-care		0.617			
Hobby		0.574			
Self-likeness					
Personality			0.722		
Praised			0.66		
Recreational activities			0.509		
Friend circle				0.866	
Shopping				0.502	
Sleep quality					0.788
Breakfast					0.551

be retained in the analysis. Thus, the stage allows reducing the dimension and interpreting the 15 variables into 5 factors.

8.5.2.3 Step 3 – Factor rotation – Interpreting factors

However, after reducing the dimension and retaining few factors it becomes difficult to interpret the criteria based on their factor loading and variances explained by the factors. For example, the first factor in Table 8.2(b) has the highest % of variance which causes most criteria to have high factor loading on them. Therefore, to address this issue, factor rotation such as Varimax with Kaiser Normalization method is conducted which allows altering the pattern of factor loadings and improving interpretation as shown in Table 8.2(c).

Hence, the rotated matrix shown in Table 8.2(c) allows the criteria to be interpreted by 5 factors such as Personal Factors, Self-care Factors, Emotional Factors, Leisure Factor and Health factors that affect the female's daily life and depression in females as shown in path model in Figure 8.2.

8.5.3 Confirmatory factor analysis – Construct validity

After the exploratory factor analysis, the study adopted confirmatory factor analysis to investigate the construct validity by testing the relationship between the 15 criteria such as Personality trait, marital status, no of children, recreational activity, weekend, self-care, hobby, breakfast, praised, sleep quality, shopping, friend circle, meeting with friends, exercise and self-likeness into to 5 factors identified as Personal Factors, Self-care Factors, Emotional Factors, Leisure Factor and

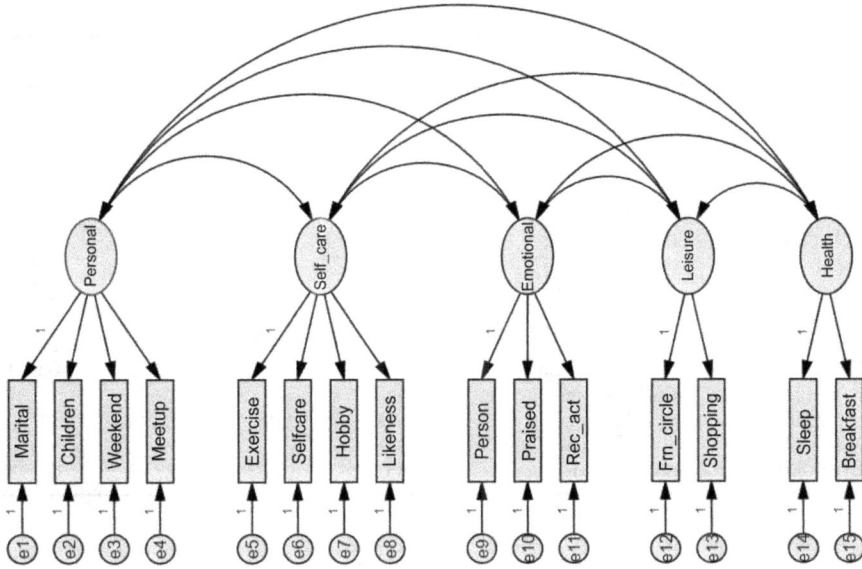

Figure 8.2 Path model

Health factors. The analysis also conducts the test to measure the goodness of model fit for the path model in Figure 8.2.

8.5.3.1 Step 1 Kaiser-Meyer-Olkin and Bartlett's test

Initially, the Kaiser-Meyer-Olkin and Bartlett's test was conducted as shown in Table 8.3. The test represents the suitability of data for structure detection. As Kaiser-Meyer-Olkin Measure of Sampling Adequacy indicates the proportion of variance of the variables that occurs due to underlying factors where a value close to 1 represents that factor analysis is the useful technique for the dataset. Finally, Bartlett's test indicates the correlation and relation of the variable where a smaller value than 0.05 indicates the significance of adopting factor analysis for the structured decisions (DiStefano and Hess, 2005).

It can therefore be concluded that the dataset is suitable for factor analysis.

Table 8.3 KMO and Bartlett's test

Kaiser-Meyer-Olkin measure of sampling Adequacy.		0.673
Bartlett's test of sphericity	Approx. Chi-Square	477.923
	df	105
	sig.	0.000

Table 8.4 Model estimates

Path			Estimate	S.E.	C.R.	p
Marital	<---	Personal	0.696			
Children	<---	Personal	0.586	0.301	5.497	***
Weekend	<---	Personal	0.800	0.202	6.734	***
Meetup	<---	Personal	0.334	0.194	3.258	0.001
Exercise	<---	Self_care	0.381			
Self-care	<---	Self_care	0.751	0.534	3.804	***
Hobby	<---	Self_care	0.511	0.300	3.363	***
Likeness	<---	Self_care	0.231	0.401	2.058	0.040
Person	<---	Emotional	0.349			
Praised	<---	Emotional	0.530	0.770	2.795	0.005
Rec_act	<---	Emotional	0.713	0.900	2.882	0.004
Frn_circle	<---	Leisure	0.440			
Shopping	<---	Leisure	0.621	0.963	3.298	***
Sleep	<---	Health	0.085			
Breakfast	<---	Health	1.852	83.618	0.257	0.798

8.5.3.2 Step 2 – Path analysis (construct validity)

The step conducts confirmatory factor analysis to investigate the construct validity as shown in Table 8.4 and Figure 8.3.

The analysis aids in testing the relationship between the criteria and the factors for example marital status, children, weekends, and meeting with friends are linked to Personal factors. It can also be are seen from Table 8.5, that all the criteria except breakfast are significant whose ($p > 0.05$), thus indicating construct validity. In addition, the analysis also provides the path estimates of the observed

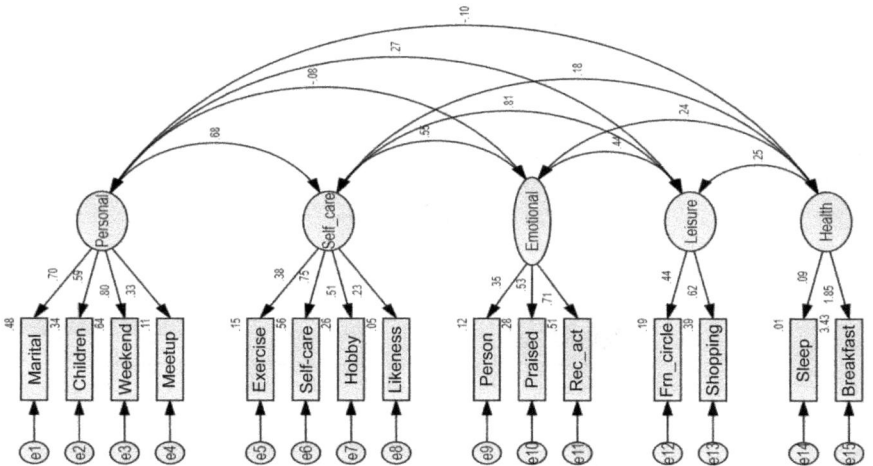

Figure 8.3 Confirmatory factor analysis path model

Table 8.5 The goodness of Model fit

Measure	Estimate	Threshold	Interpretation
CMIN	214	–	–
DF	80	–	–
CMIN/DF	2.6	Between 1 and 3	Acceptable
CFI	0.662	>0.95	Not acceptable
RMSEA	0.117	<0.06	Not acceptable

criteria that represent the correlation coefficients between them and the factor that represents them is shown in Figure 8.3.

8.5.3.3 Step 3 – Goodness of fit

Furthermore, to determine whether the path model is acceptable, the goodness of fit indices such as CMIN/DFI (minimum discrepancy per degree of freedom), CFI (comparative fit index), and RMSE (Root Mean Square Error) as suggested by Mulaik (1989) were adopted by the study. The goodness of fit indices helps to summarize the inconsistency between observed values and the values expected under the model in question. Table 8.5 shows the estimated values of indices calculated, ideal threshold values, and interpretation of these indices.

For the model to be acceptable must perform well and within the threshold values. Although the CMIN/DF indicates the acceptable goodness of fit. However, the estimates of CFI and RMSE indicate a poorly specified model. Thus, in an event like this, it is recommended that the researchers make the necessary adjustment in the model by adjusting the survey, collecting more observations, and fitting different models (Lance, Butts, and Lawrence, 2006).

8.6 Conclusion

Depression being one of the 10th leading causes of early death and loss of productivity is a critical factor. Literature reveals the dynamic nature of depression and a huge number of variables that contribute to the mental health and wellbeing of society in general and females in particular. As such, some studies have identified 18 variables that relate to the personal and social aspects of women's daily life which contribute to their depression. However, due to the unknown relationship between these criteria and their impact on women's depression, this study adopts exploratory and confirmatory factor analysis to investigate the construct validity and relationship between the criteria and factors. The analysis aid in interpreting the 18 criteria into five factors as Personal factors, self-care factors, Emotional Factors, Leisure Factors and Health factors that affect the female's life and depression in females. This study attempted to develop a model that represents the relationship between these criteria and the defined factors. The goodness of fit indices such as CFI (comparative fit index) and RMSE (Root Mean Square Error) showed that this model is not fit, while the CMIN/DFI (minimum discrepancy per degree of freedom) test showed acceptable goodness of fit. Thus, to improve the goodness of fit the study recommend fitting different model, i.e. retaining a different number of factors and by using different rotation method. The approach adopted by this study (with successful goodness of fit tests) will help the study with developing a structural equation modeling.

References

Atkinson, T.M., Rosenfeld, B.D., Sit, L., Mendoza, T.R., Fruscione, M., Lavene, D., Shaw, M., Li, Y., Hay, J., Cleeland, C.S. and Scher, H.I., 2011. Using confirmatory factor analysis to evaluate construct validity of the Brief Pain Inventory (BPI). *Journal of Pain and Symptom Management*, 41(3), pp. 558–565.

Brown, J.D. 2000. Statistics Corner: What is Construct Validity? Shiken: JALT Testing & Evaluation SIG Newsletter, 4(2) Oct. 2000 (p. 8–12) http://hosted.jalt.org/test/PDF/Brown8.pdf

DiStefano, C. and Hess, B., 2005. Using confirmatory factor analysis for construct validation: An empirical review. *Journal of Psychoeducational Assessment*, 23(3), pp. 225–241.

Fairgaze., 2021. Role of Women in Society. Available at: https://fairgaze.com/interested-article/role-of-women-in-society.htm

Habing, B., 2003. Exploratory factor analysis theory and application. In [Online], [Retrieved April 22, 2020], Available: https://www.let. rug. nl/nerbonne/teach/rem a-stats-meth-seminar/Factor-Analysis-Kootstra-04.PDF.

Jabeen, H., 2014. Adapting the built environment: The role of gender in shaping vulnerability and resilience to climate extremes in Dhaka. *Environment and Urbanization*, 26(1), pp. 147–165.

Jowit, J., 2018. "What is Depression and why it is rising?". *Guardian News*. Available Online: https://www.theguardian.com/news/2018/jun/04/what-is-depression-and-why-is-it-rising.

Kim, J.; Lee, J., 2013. Prospective study on the reciprocal relationship between intimate partner violence and depression among women in Korea. *Social Science & Medicine*, 99, p. 42. ISSN 0277-9536.

Levant, R.F., Rogers, B.K., Cruickshank, B., Rankin, T.J., Kurtz, B.A., Rummell, C.M., Williams, C.M. and Colbow, A.J., 2012. Exploratory factor analysis and construct validity of the Male Role Norms Inventory-Adolescent-revised (MRNI-Ar). *Psychology of Men & Masculinity*, 13(4), p. 354.

Lance, C.E., Butts, M.M. and Michels, L.C., 2006. The sources of four commonly reported cutoff criteria: What did they really say? *Organizational Research Methods*, 9(2), pp. 202–220.

Lombardo, C. et al., 2014. Severity of insomnia, disordered eating symptoms, and depression in female university students. *Clinical Psychologist*, 18(3), pp. 108–115. ISSN 1328-4207.

McIlvenny, S., DeGlume, A.M., Elewa, M., Fernandez, O.T. and Dormer, P., 2000. Factors associated with fatigue in a family medicine clinic in the United Arab Emirates. *Family Practice*, 17(5), pp. 408–413.

Mulaik, S.A., James, L.R., Van Alstine, J., Bennett, N., Lind, S. and Stilwell, C.D., 1989. Evaluation of goodness-of-fit indices for structural equation models. *Psychological Bulletin*, 105(3), p. 430.

Hussain, F.; Cochrane, R., 2002. Depression in South Asian women: Asian women's beliefs on causes and cures. *Mental Health, Religion & Culture*, 5(3), pp. 285–311. ISSN 1367–4676.

Piccinelli, M. and Wilkinson, G., 2000. Gender differences in depression: Critical review. *The British Journal of Psychiatry*, 177(6), pp. 486–492.

Razzak, H.A., Harbi, A. and Ahli, S., 2019. Depression: Prevalence and associated risk factors in the United Arab Emirates. *Oman Medical Journal*, 34(4), p. 274.

Sehlo, M. G.; Bahlas, S. M., 2013. Perceived illness stigma is associated with depression in female patients with systemic lupus erythematosus. *Journal of Psychosomatic Research*, 74(3), pp. 248–251. ISSN 0022-3999.

Regestein, Q., Natarajan, V., Pavlova, M., Kawasaki, S., Gleason, R. and Koff, E., 2010. Sleep debt and depression in female college students. *Psychiatry Research*, 176(1), pp. 34–39. DOI: 10.1016/j.psychres.2008.11.006

Sung, C.-W. et al., 2016. Early dysautonomia detected by heart rate variability predicts late depression in female patients following mild traumatic brain injury. *Psychophysiology*, 53(4), pp. 455–464. ISSN 0048-5772.

Tong, J. et al., 2014. A two-stage epidemiologic study on prevalence of eating disorders in female university students in Wuhan, China. *Social Psychiatry and Psychiatric Epidemiology*, 49(3), pp. 499–505. ISSN 0933-7954.

WHO., 2021. Depression. Available at: https://www.who.int/news-room/fact-sheets/detail/depression

Williams, B., Onsman, A. and Brown, T., 2010. Exploratory factor analysis: A five-step guide for novices. *Australasian Journal of Paramedicine*, 8(3) pp 1–13.

Women, U.N., 2019. Progress on the Sustainable Development Goals: The gender snapshot 2019.

Part IV

Other research reliability and validity approaches

9 A case study of Singapore's hawker centres as an inclusive mechanism

Internal and external validity of qualitative data

Yajian Zhang and Willie Tan

9.1 Background

Inclusive development has received much attention over the last two decades to address the growing problems of socioeconomic inequalities and spatial isolation because of globalization and fiscal crises (Friend et al., 2016, McGranahan et al., 2016, Shrestha et al., 2014). Many studies have examined inclusive mechanisms from four main perspectives. The poor may be excluded in various ways, such as spatially (Haughton et al., 2009), socially (Dempsey et al., 2011), economically (Shortall, 2004) and politically (Young, 2002). These four dimensions of exclusion are related and tend to interact. For example, many cities consist of different political, economic and social areas.

Historically, towns and cities were exclusive places (Sjoberg, 1960). In modern economies, exclusionary mechanisms have been used or are still in use, especially in housing markets. These devices include:

1 steering particular races to certain neighbourhoods (Galster, 1990);
2 fiscal and exclusionary zoning (Pogodzinski, 1991) to raise tax revenues and exclude the poor from consuming public goods and services in wealthier neighbourhoods;
3 mortgage lenders' redlining of certain parts of town as high-risk of default areas (Holmes and Horvitz, 1994);
4 gated communities to keep non-residents out of the local development (Blakely and Snyder, 1997); and
5 privatization and commercialization of public places (Mitchell, 1995).

In the developing countries, economic exclusion takes the primary form of informal employment and housing in large cities where marginalized migrants work and live in slum-like conditions. These mechanisms reflect the concerns of urban elites regarding excessive urbanization, such as the apartheid system in South Africa (Maylam, 2017), and the household registration (hukou) system in China (Chan and Zhang, 1999).

DOI: 10.1201/9780429243226-13

The concentration of the lower classes in unhealthy slums is a recipe for disaster. In the 1980s, urban unrests broke out because of the global debt crises and IMF's "structural adjustment programs". Subsequently, the IMF softened its stand with its "adjustment with a human face" programme to protect the vulnerable (Cornia et al., 1987). The World Bank also began to promote "good governance" rather than just "getting prices right" as part of economic development. A component of good governance is inclusive development. Since then, many inclusive mechanisms have been adopted, such as participatory planning, creation of mixed-use communities, easy access to physical and social infrastructure, creating job opportunities, inclusive urban design and ending racial and religious discrimination (Burton and Mitchell, 2006, Espino, 2015, Godschalk, 2004, Hambleton, 2014, Williams and Pocock, 2010).

In this chapter, we use a narrative to investigate how Singapore uses its hawker centres to promote inclusive development. We identify how hawker centres originated, how they developed over time and the outcomes. The chapter consists of five sections. Section 2 provides the research methodology, followed by the case study in Section 3. Section 4 tests the validity and reliability of the findings. The final section concludes the chapter.

9.2 Research methodology

The research methodology is presented in Table 9.1.

Table 9.1 Research methodology for this study

Research design	Choice	Reason
Philosophy	Realist	This study assumes the existence of a real object of inquiry, and we can discover its properties through theory and evidence. Even though data are theory-laden, they are not theory-determined.
Approach	Inductivist	The study is exploratory, and proceeds from the ground, rather than deductive testing of prior hypothesis. It uses a framework to guide the study, based on the proposition that political struggles lie at the centre in hawker centre development in Singapore.
Strategy	Case study	A case study is adopted to provide an in-depth investigation of hawker centres as they developed out of struggles among the main stakeholders.
Methods	Qualitative	This study uses a narrative to tell the story of how hawker centres developed as an inclusive mechanism in Singapore.
Tools	Analysis of past documents	This study makes extensive use of secondary data to trace the historical development of hawker centres in Singapore.

9.3 Results and main findings

9.3.1 *Origin of concept of hawker centres in Singapore*

In 1819, Raffles negotiated with the local Malay ruler to establish the British colony of Singapore as a free port. The European agency houses dominated the trade, and they worked with predominantly Chinese compradors who were knowledgeable about regional Southeast Asian networks. The new port city attracted many immigrants but, without industrialization, employment opportunities were limited and street hawking grew along with trade and financial services. These hawkers usually lacked proper equipment, water supply and waste disposal system, which created concerns with food hygiene and environmental pollution. Additionally, street hawking also affected vehicular and pedestrian traffic (Kong, 2007).

During the early 20th century, the city had become overcrowded and the colonial government began to intervene more extensively in the built environment. Its first priority was housing and, in 1927, it established the Singapore Improvement Trust (SIT) to provide public rental housing. Next, in 1931, the colonial government formed the Hawker Advisory Committee to address the hawking issue. According to the *Report of the Committee Appointed to Investigate the Hawker Question in Singapore* (1932), the colonial government's initial focus was on regulating eating and coffee shops. In 1913, the municipal health authorities imposed high standards on such shops through a licensing scheme, resulting in a sharp decline from 1,520 shops in 1914 to only 515 shops in 1922. Consequently, the number of street hawkers grew. Licensing was implemented from 1919 and by 1931, there were 211 day stalls, 462 night stalls and 6,042 itinerant hawkers that were licensed but only six shelters for 383 hawkers and about 4,000 unlicensed street hawkers.

The dispersed nature of illegal hawking made enforcement difficult. The unlicensed hawkers tended to avoid police raids by playing cat and mouse, bribing the police, or paying protection fees to gangsters. The Committee recommended reducing the number of itinerant hawkers, narrowing the areas they could trade for better control and building new markets for licensed hawkers. However, the Second World War disrupted the implementation of the recommendations.

By the 1950s, only about one-third of the hawkers had been licensed (Grice, 1988). Additionally, only two new hawker markets were completed from 1931 to 1950 because of other post-Depression and war priorities. Tensions grew as the police tried to arrest unlicensed hawkers and destroy their equipment and stock. In 1948, political events took a decisive turn when the Communist Party of Malaya (CPM) took up an armed struggle to agitate for an independent Malaya and destroyed many rubber plantations and mines. Shortly after, it infiltrated the urban areas, particularly working class and student associations (Short, 1975).

In 1950, a Hawker Inquiry Commission was formed to ease the tension. The Commission organized a public meeting and obtained feedback from the municipality, Hawkers Association and the public. The Commission acknowledged the benefits of cheap and convenient hawker food and recommended the building of hawker centres in residential areas for better control of hawkers. Rents were set to be competitive at close to the licensing fee charged on street hawkers to encourage hawkers to take up the offers. There were also efforts from the private sector. Some hawkers formed a syndicate to buy a piece of land and build their own market. In 1954, a private entrepreneur built a market in Serangoon Road and rented the stalls to hawkers.

Despite these initial attempts, there were too few hawker centres to resettle the large number of illegal hawkers. These hawkers also enraged coffee shop owners by selling food near their shops, thereby competing with their renter stallholders. In January 1953, the Kheng-chew and Foo-chew Coffee Shop Owners Associations with more than 2,000 members petitioned against hawkers selling food near their shops. The Singapore Hawkers Association countered that approval of this request would deprive many hawkers of their livelihoods (The Singapore Free Press, 1953). Although the colonial government established the Markets and Hawkers Department in 1957 to oversee the trade, the matter was unresolved and illegal hawking was still prevalent.

9.3.2 A new strategy

In 1959, Singapore achieved self-government and the new government embarked on a new development path (*State of Singapore Development Plan, 1961–1964*). The broad strategy was to industrialize using private and foreign investment, provision of infrastructure and industrial estates, technical training, labour control and tax incentives. Importantly, it disbanded the City Council to control the provision of municipal services. Hence, it established the Housing and Development Board (HDB) in 1960 to provide mass public housing and build commercial, neighbourhood and hawker centres in its housing estates. Outside these suburban residential areas, hawker centres were also built in the inner city areas and industrial states by the Urban Redevelopment Authority (URA) and Jurong Town Corporation (JTC), respectively. Above all, in the Concept Plan of 1971, Singapore was to become a new Garden City in its, not a city of slums.

With these new hawker centres at subsidized rents in place or under construction, the government began to intervene directly into the hawking trade to clear the streets of illegal hawkers in the Garden City through registration, relocation and more effective enforcement. By 1973, there were approximately 38,000 registered hawkers. The government formed a special squad and empowered the public health inspectors to crack down on illegal hawking. Severe penalties were imposed on illegal hawkers caught plying their trade.

The hawker centre building programme was a success. From 1971 to 1985, 135 modern hawker centres were built (Tay, 2014). By 1986, the government had resettled all street hawkers. Consequently, it stopped building new hawker centres. Figure 9.1 provides an example of a hawker centre located in Clementi, built in 1980.

Hawker centres were not just eating places. It was also a place for people from different races and backgrounds to mingle and bond, and to participate in various activities such as night markets, religious festivals and food festivals. A survey conducted by NEA and Ministry of Community Development in 2005 indicated that 81% of respondents agreed that hawker centres played an important role in community bonding (Kong, 2007).

Figure 9.1 A hawker centre in Clementi, Singapore

Source: Authors

After 1986, the government built many coffee shops in HDB estates to ensure the sufficient supply of affordable food. These coffee shops are located at the ground floor of residential buildings. Unlike hawker centres that are large food establishments comprising 30 to 200 stalls, a coffee shop (or coffee house) has about 10 stalls that sell coffee, local dishes and other beverages (Figure 9.2).

In 2001, the government started to upgrade hawker centres to better provide services and serve as community spaces for social interaction. Some of them were completely reconfigured or rebuilt. In 2011, the government announced that it would build 10 new hawker centres by 2027. The new supply would reduce rents and moderate food prices. Many of these centres will be located in outlying new towns to serve residents. Additionally, social enterprises were encouraged to operate and manage hawker centres.

9.3.2.1 Outcomes

Currently, there are 114 hawker centres in Singapore (Table 9.2). The National Environment Agency (NEA) manages 107 hawker centres owned by the government and HDB. Seven were managed by social enterprises.

The number of hawker stalls showed a steady decline from 1989 to 2003 (Figure 9.3) largely from retirements. There was a sharp increase in 2004 because the completed upgrade programme in 2003 created many extra stalls.

For the licensing of hawkers, Singapore citizens or permanent residents are eligible for application. Table 9.3 presents the annual number of licensed hawkers

Figure 9.2 A coffee shop in Singapore
Source: Authors

from 2013 to 2017. There were 13,865 licensed hawkers in 2017. About half of them (5,999) sold cooked food in hawker centres, and a small percentage of licenses (536) were issued for street hawkers (Figure 9.4).

9.4 Reliability and validity

9.4.1 Reliability

Reliability refers to the accuracy and precision of measurement. In the quantitative context, measures that are accurate and with smaller standard errors are more reliable. Accuracy is a measure of correspondence to the truth, or the real object. In qualitative studies, reliability also refers to the authenticity of data, rather than the measures themselves.

Table 9.2 Hawker centres in Singapore, 2019

Ownership	Number
Government	80
Housing Development Board (HDB)	27
Social enterprises	7
Total	114

Source: NEA.

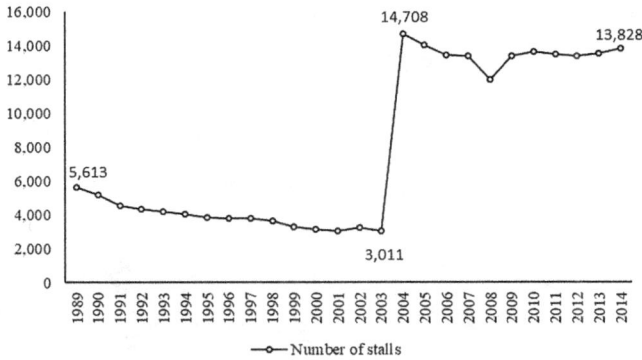

Annual number of hawker stalls

Value	Label
16,000	
14,000	14,708
12,000	13,828
10,000	
8,000	
6,000	5,613
4,000	
2,000	3,011
0	

—o— Number of stalls

Figure 9.3 Annual number of hawker stalls from 1989 to 2014

Source: data.gov.sg, redrawn by Authors

The reliability of data sources can be ascertained through source criticism and consistency with other texts. For source criticism, we examined various sources and relied largely on official data. These data are cross-referenced with other texts, such as scholarly books, newspaper reports and journal articles to ensure that the numbers, events and interpretations are consistent. Finally, we also use images to illustrate the concept of hawker centres, coffee shops and street hawking.

9.4.2 Internal validity

In qualitative studies, internal validity refers to correspondence between concepts and measures, and theoretical consistency linking causes and effects (Tan, 2017).

For the correspondence between concepts and measures, we need to be careful about the key concepts, in this case, hawkers and inclusiveness. Here, a hawker is defined as a stallholder who sells food and drinks. A hawker may be rich or poor, or may rent or own his/her stall. The type or location of stall does not matter, or whether a hawker is plying his trade legally or illegally. For inclusiveness, we rely on the commonly accepted dimensions of political, social, spatial and

Table 9.3 Annual number of licensed hawkers from 2013 to 2017

Year	2013	2014	2015	2016	2017
Total licensed issued	14,227	14,466	14,055	13,871	13,865
Type of premises					
Hawker centres	13,537	13,828	13,440	13,310	13,329
Street	690	638	615	561	536
Type of goods sold					
Market produce	5,706	5,857	5,612	5,485	5,479
Cooked food	5,939	6,049	5,943	5,970	5,999
Piece and sundries	2,582	2,560	2,500	2,416	2,387

Source: Department of Statistics, Singapore.

Figure 9.4 A street stall in Chinatown, Singapore

Source: Authors

economic inclusiveness. For example, in the case of political inclusiveness, we show how the Hawkers Inquiry Commission of 1950 received public feedback from the Hawkers Association. More generally, Singapore's multi-racial society makes the government more sensitive to feedback to avoid the racial riots and disharmony of the past.

For the concept of spatial inclusiveness, how does one select proper sites for hawker centres? The street hawkers prefer to relocate to stalls near their original sites to mitigate the risk of losing regular customers. As most street hawkers operated in the central urban area before the 1970s, it was impossible to relocate all of them to new hawker centres in the city centre. The Markets and Hawkers Department worked with hawkers to understand their needs and minimized the disruption to business. The final solution, based on Singapore's new development strategy, was to build hawker centres in public housing estates within walking distance of residents.

In Singapore, public housing estates are not just homes. They also provide an opportunity to build communities. For this reason, there is a quota system of public housing allocation to avoid dense racial concentration in public housing estates. To reduce perceptions of economic disparity, HDB mixes its public housing estates by housing types (3, 4 and 5-room flats) and about 20% are private housing. Various grants are also available to encourage extended families to live near each other. In this context, it is easy to see why hawker centres serve as a space for people of different races and incomes to intact and have good and affordable food.

Table 9.4 Subsidized rent of hawker stalls

Types of goods sold	Subsidized rent per month (Singapore $)		
	Without upgrading	*Standard upgrading*	*Reconfiguration/ rebuilding*
Cooked food	160	192	320
Piece and sundries	92	110.40	184
Market products I	80	96	160
Market products II	56	67.20	112

Source: Tan (2015).

Finally, for economic inclusiveness, stalls in hawker centres were initially allocated to previous street hawkers or displaced hawkers. As more hawker centres were built and some hawkers retired or gave up the business, there were vacant stalls. In 1975, under its hardship scheme, the government began to rent vacant stalls at subsidized rate to poor people above 40 years of age with monthly family income of less than $500. Since every hawker would request for a good location, the government used a ballot system to ensure fairness and transparency. However, this system could not ensure food variety because hawkers selling the same products could be located side by side. Some approved applicants declined stalls allocated to them because they were not satisfied with the locations. In 1990, NEA replaced the balloting system with an open monthly tender scheme to address these two problems. The tender system also reduces the rent subsidy and hence promotes greater efficiency through competition.

Currently, there are subsidized and non-subsidized stalls. Subsidized stallholders are the original street hawkers and those who meet the criteria of the hardship scheme. They can enjoy subsidized rents (see Table 9.4) as long as they are not in other occupations and operate their stalls personally.

For non-subsidized stalls, rents are determined by the market. According to NEA, average successful tender bid for cooked food stalls from 2015 to 2018 was $1,514 per month. Rents of about 85% of cooked food stalls were below $1,500. The highest and lowest bids for hawker stalls from April 2018 to March 2019 are given in Figure 9.5.

NEA has adopted several measures to moderate the rents of hawker stalls. In 2012, it abolished the reserve rental price and published successful bids on its website to assist tenderers in their bids. Non-subsidized stallholders cannot sublet or assign their stalls. However, a subsidized stallholder who wishes to leave the business could assign the stall. The assignee's rent will be progressively raised every year from the subsidized to assessed market rent over three years.

The government provides the Hawkers Productivity Grant to encourage hawkers to automate. Eligible stallholders can claim 80% of the equipment cost with a cap of $5,000 over three years. In view of the large number of retiring hawkers, the government introduced the Incubation Stall Programme to encourage young hawkers to join the business. Successful applicants will enjoy a discounted rent at

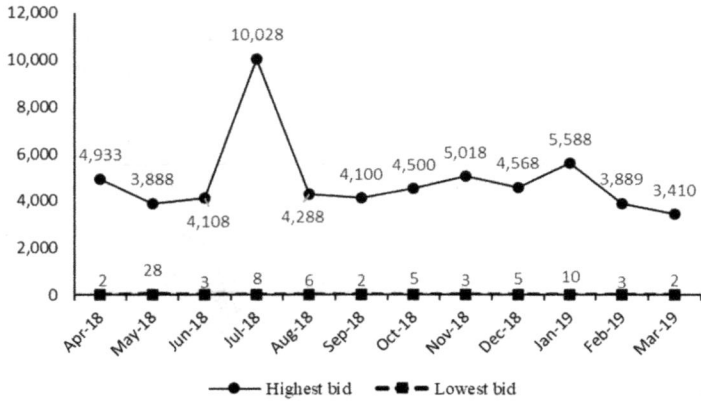

Figure 9.5 Bids for hawker stalls from April 2018 to March 2019

Source: NEA, redrawn by Authors

50% of the market price for nine months. The stalls are also equipped with basic facilities to lower the start-up cost.

The final component of internal validity is theoretical consistency linking causes and effects. This consists of two parts, namely, the framework to guide the narrative and the causal mechanism. For the framework, we choose to place politics at the centre of hawker centre development because, in Singapore's context, the development of hawker centres is not based on private enterprise. It is a product of the Developmental State as an alternative to the neoliberal free market approach to economic development (Amsden, 1994, Wade, 1990).

For the causal mechanism, we need to show, for example, why the government intervened directly into hawker centres. It was part of a new economic development strategy to provide employment for the poor, clean the streets of Singapore to build a new Garden City, and build social communities in housing estates.

9.4.3 External validity

External validity refers to the generalizability of its findings. Hawker centres are found in many parts of Southeast Asia. The key question here is that of implementation and inclusiveness so that hawker centres can serve as a model to reduce broader issues of poverty.

We have shown that the concept of hawker centres in Singapore is not a post-colonial one; it originated with the colonial government. Where it differs in Singapore is the direct intervention of the post-colonial government to transform and restructure its urban form to facilitate a new industrial future. While not all States have the capacity to mobilize the resources, it is also true that many small gains to help the poor can be achieved through limited reforms, or what is known as "good enough governance" (Grindle, 2004). There is no point in waiting for a long list of good governance reforms to take place before taking action to alleviate poverty.

We have also shown that direct State intervention is not the only way to build hawker centres. During colonial times, we provided two examples, namely, self-built and managed hawker centres and privately built hawker centres. More recently, the government has been encouraging social enterprises and young Singaporeans to enter the hawker business.

9.5 Conclusions

In this chapter, we make extensive use of secondary data to investigate how Singapore uses its hawker centres to promote inclusive development. We use a political framework to provide a coherent narrative on the origin of hawker centres, their development and the outcomes. Our main findings are that hawker centre development depends very much on the priorities of governments, which in turn hinges on historical struggles between hawkers and other stakeholders, or what Scott calls the "weapons of the weak" (Scott, 2008). These resistances, at times violent, are structurally mediated by the State as a neutral referee or an interested stakeholder; at times, these conflicts were ignored and tolerated; at other times, they became instruments of various forms of inclusiveness and economic development. Finally, we provide extensive commentaries and examples on the reliability and validity of this study.

References

Amsden, A.H. (1994). Why isn't the whole world experimenting with the East Asian model to develop?: Review of the East Asian miracle. *World Development*, 22, 627–633.

Blakely, E.J. & Snyder, M.G. (1997). *Fortress America: Gated Communities in the United States*, Washington, D.C.: Brookings Institution Press.

Burton, E. & Mitchell, L. (2006). *Inclusive Urban Design: Streets for Life*, London: Routledge.

Chan, K.W. & Zhang, L. (1999). The *hukou* system and rural-urban migration in China: Processes and changes. *The China Quarterly*, 160, 818–855.

Committee Appointed to Investigate the Hawker Question in Singapore (1932). *Report of the Committee Appointed to Investigate the Hawker Question in Singapore*, Singapore: Government Printing Office.

Cornia, G., Jolly, R. & Stewart, F. (eds.) (1987). *Adjustment with a Human Face: Volume 1, Protecting the Vulnerable and Promoting Growth*, Gloucestershire: Clarendon Press.

Dempsey, N., Bramley, G., Power, S. & Brown, C. (2011). The social dimension of sustainable development: Defining urban social sustainability. *Sustainable Development*, 19, 289–300.

Espino, N.A. (2015). *Building the Inclusive City: Theory and Practice for Confronting Urban Segregation*, London: Routledge.

Friend, R.M., Anwar, N.H., Dixit, A., Hutanuwatr, K., Jayaraman, T., McGregor, J.A., Menon, M.R., Moench, M., Pelling, M. & Roberts, D. (2016). Re-imagining inclusive urban futures for transformation. *Current Opinion in Environmental Sustainability*, 20, 67–72.

Galster, G. (1990). Racial steering by real estate agents: Mechanisms and motives. *The Review of Black Political Economy*, 19, 39.

Godschalk, D.R. (2004). Land use planning challenges: Coping with conflicts in visions of sustainable development and livable communities. *Journal of the American Planning Association*, 70, 5–13.

Grice, K. (1988). *The institutionalisation of informal sector activities: a case study of cooked food hawkers in Singapore*. PhD Thesis, Keele University, Staffordshire, United Kingdom.

Grindle, M.S. (2004). Good enough governance: poverty reduction and reform in developing countries. *Governance*, 17, 525–548.

Hambleton, R. (2014). *Leading the Inclusive City: Place-based Innovation for a Bounded Planet*, Bristol, UK: Policy Press.

Haughton, G., Allmendinger, P., Counsell, D. & Vigar, G. (2009). *The New Spatial Planning: Territorial Management with Soft Spaces and Fuzzy Boundaries*, London: Routledge.

Holmes, A. & Horvitz, P. (1994). Mortgage redlining: Race, risk, and demand. *The Journal of Finance*, 49, 81–99.

Kong, L. (2007). *Singapore Hawker Centres: People, Places, Food*, Singapore: National Environment Agency.

Maylam, P. (2017). *South Africa's Racial Past: The History and Historiography of Racism, Segregation, and Apartheid*, London: Routledge.

McGranahan, G., Schensul, D. & Singh, G. (2016). Inclusive urbanization: Can the 2030 Agenda be delivered without it? *Environment and Urbanization*, 28, 13–34.

Mitchell, D. (1995). The end of public space? People's Park, definitions of the public, and democracy. *Annals of the Association of American Geographers*, 85, 108–133.

Pogodzinski, J.M. (1991). The effects of fiscal and exclusionary zoning on household location: A critical review. *Journal of Housing Research*, 2, 145–160.

Scott, J.C. (2008). *Weapons of the Weak: Everyday Forms of Peasant Resistance*, New Haven: Yale University Press.

Short, A. (1975). *The Communist Insurrection in Malaya, 1948–1960*, New York: Crane, Russak.

Shortall, S. (2004). Social or economic goals, civic inclusion or exclusion? An analysis of rural development theory and practice. *Sociologia Ruralis*, 44, 109–123.

Shrestha, K.K., Ojha, H.R., McManus, P., Rubbo, A. & Dhote, K.K. (eds.) (2014). *Inclusive Urbanization: Rethinking Policy, Practice and Research in the Age of Climate Change*, New York: Routledge.

Singapore Department of Statistics. *Licensed Hawkers Under National Environment Agency (End of Period), Annual* [Online]. Available: https://www.tablebuilder.singstat.gov.sg/publicfacing/createDataTable.action?refId=14624 [Accessed 08 May 2019].

Singapore National Environment Agency (NEA). *Markets/Hawker Centres Managed by NEA* [Online]. Available: https://www.nea.gov.sg/docs/default-source/our-services/hawker-management/list-of-hc.pdf [Accessed 28 March 2019].

Singapore National Environment Agency (NEA). *NEA Adopts Transparent Tender System for Hawker Stalls (6 Sep 2018)* [Online]. Available: https://www.nea.gov.sg/media/readers-letters/lists/replies-issued-by-nea/nea-adopts-transparent-tender-system-for-hawker-stalls-(6-sep-2018) [Accessed 27 April 2019].

Singapore National Environment Agency (NEA). *Successful Tenders* [Online]. Available: https://www.nea.gov.sg/corporate-functions/resources/tender-notices [Accessed 15 May 2019].

Sjoberg, G. (1960). *The Preindustrial City: Past and Present*, Glencoe, Illinois: Free Press.

Tan, S.B. (2015). *Keeping Char Kway Teow Cheap - At What Price?*, Singapore: National University of Singapore.

Tan, W. (2017). *Research Methods: A Practical Guide for Students and Researchers*, Singapore: World Scientific.

Tay, R. (2014). *Hawker Centres: Levelling the Playing Field with Food*, Singapore: Centre for Liveable Cities.

The Singapore Free Press. (1953). They Fear These Hawkers. *The Singapore Free Press*, January 1953, p. 5.

Wade, R. (1990). *Governing the Market: Economic Theory and the Role of the Government in East Asian Industrialisation*, New Jersey: Princeton University Press.

Williams, P. & Pocock, B. (2010). Building 'community'for different stages of life: physical and social infrastructure in master planned communities. *Community, Work & Family*, 13, 71–87.

Young, I.M. (2002). *Inclusion and Democracy*, Oxford: Oxford University Press.

10 The Delphi technique as a tool for quality research in the built environment

Edoghogho Ogbeifun and Jan-Harm C. Pretorius

10.1 Introduction

The strength or weakness of any research exercise depends on the quality of data, the sources and the instrument(s) used for data collection, as well as the method(s) of analysis. To improve on the reliability and credibility of research exercises, there is a continuous attempt at improving the methods of data management. The majority of the suggested improvements include the reduction in the influence or interference and bias of the researcher. In this regard, the procedure of executing a typical Delphi technique provides that the researcher, commonly referred to as the coordinator, facilitates and allows the participants to run and decide the outcome of the research. The two case studies discussed in this chapter focus on unique problems within the context of the sites chosen for the research. By adopting, in part, the principles of participatory research, the participants who are highly knowledgeable persons in the subject of the research, were involved in finding solutions to the research problems.

The problems in the case studies border on operational efficiencies in the provision of functional facilities for the performance of the core functions of teaching, learning and research within suitable academic environments. The Delphi technique was adopted as the research method against the philosophy that "several people are less likely to arrive at a wrong decision than a single individual". The technique was used as instrument for data collection and the execution of the entire research exercise.

The outcome of the two research exercises suggests that the respective stakeholders were satisfied with the results. In the first exercise, the client and the service provider (FM unit) accepted the developed KPIs as useful tools which will help improve on the performance of the FM unit in the provision, operation and maintenance of the required facilities for the academics to perform their core functions of teaching, learning and research. Similarly, identifying the external and internal factors responsible for the delays in the execution of construction projects in the second case study armed the operators with useful information on what to do in order to ameliorate the incidence of delays.

The validity and reliability of the Delphi technique as a research tool was achieved by selecting a panel of experts who are knowledgeable in the subject

DOI: 10.1201/9780429243226-14

of the research and met the defined pre-qualification criteria. The results of the two research exercises were refined progressively from one round to another, until consensus was achieved. Although, some of the participants knew each other, the individual integrity and anonymity of opinions were not compromised. At the end of the exercise, participants were satisfied with the outcome.

10.2 Literature review

The discussion on the question of validity and reliability in any research exercise essentially means aiming the search light onto the quality of data, sources of data, the tool(s) used for data collection and the methods of data analysis. To achieve validity and reliability, different research methods are available; some of them can be used as stand-alone methods and some are used in conjunction with other tools. In the built environment research, the Delphi technique is becoming a commonly accepted technique for quality research (Hallowell & Gambatese, 2010; Hallowell *et al.*, 2013; Agumba *et al.*, 2014; Musonda & Pretorius, 2015; Alaloul *et al.*, 2015; Kermanshachi *et al.*, 2016). The strength or reliability of the technique centers on the quality of the participants, known as experts or highly knowledgeable persons in the research area (Day & Bobeva, 2005; Grisham, 2009). The participants are purposively selected, not limited by geographical location and they can be few in number or as many as possible. The process requires participants who are willing to apply themselves to the repetitive cycles of the data collection (Day & Bobeva, 2005). Due to the repetitive nature of the Delphi process, some participants do drop out in between the different rounds. Therefore, it is necessary to recruit high number of participants at the beginning of the exercise in order to manage the negative effects of possible attrition (Donohoe & Needham, 2009; Adnan & Daud, 2010). Furthermore, the Delphi process includes ensuring the anonymity of participants' contributions and results being refined in every cycle of data collection. Participants adjust their position into the group's adjudged superior opinion, without coercion, until equilibrium is achieved (Ogbeifun *et al.*, 2016).

The Delphi technique may not fit perfectly into the classical divide between qualitative and quantitative research methodologies, but as a "hybrid" research method (Sekayi & Kennedy, 2017). It is a technique that integrates the elements of both qualitative and quantitative methodologies in a single research exercise. It is one of the few research tools which allows participants to respond to the same research exercise repeatedly and view the contributions of other participants. There are different variants of the Delphi technique. The two used in this paper are the classic and the modified classic Delphi techniques. The difference between the two variants is that in the classic Delphi, participants generate the prospective solutions to the research question in the first round of data collection. While in the modified classic Delphi, participants are provided with generic solutions to the research question in the first round (Franklin & Hart, 2007). In subsequent rounds, participants score the proposed solutions within a defined scale of measure. After every iteration, only items that scored above the benchmark set by the group or the coordinator are escalated to the next round. The Delphi technique can be used at the beginning to kick start a research process or at the end of the research process to fine-tune results obtained from other research methods (Day & Bobeva, 2005; Hasson & Keeney, 2011, Habibi *et al.*, 2014). In the Delphi technique, consensus is reached after continuous recycling of results among participants until equilibrium of opinion is achieved. Equilibrium is achieved when participants no longer change their opinions or the process achieves between 51% and 80% agreement in the list of items in the final round (Hasson *et al.*, 2000; Hasson & Keeney, 2011). This can be achieved in few or many rounds of successive iterations. The controlled feedbacks allow participants to view their individual submissions in the light of the whole group, which may result in the adjustment of an individual's opinion without coercion (Hasson & Keeney, 2011; Ogbeifun *et al.*, 2017).

This research tool has been used by these authors in many studies of which two case studies are discussed in this paper. Firstly, it was used for the development of key performance indicators (KPIs) for measuring the performance of the operation of the Facilities Management (FM) unit in a higher education (HE) institution in South Africa. The second was a pilot study aimed at exploring the causes of delays in the execution of a construction project, funded by an agency of the Federal government of Nigeria where there is evidence of adequate funding.

10.3   Research methodology

The research reports being presented in this chapter adopted the case study method of qualitative research, because "case studies explore and investigate contemporary real-life phenomenon[s] through detailed contextual analysis of a limited number of events or conditions, and their relationships" (Zainal, 2007, p. 1). Further, "[a] case study is an intensive study of a single unit for the purpose of understanding a larger class of (similar) units" (Gerring, 2004, p. 342). This method is useful when holistic, in-depth investigation is needed (Green & Thorogood, 2009; Lateef *et al.*, 2010). The first case study adopted the single site, while the second adopted the multiple sites case study formats (Holliday, 2007; Yin, 2014).

The Delphi technique was used as an instrument for data collection and analysis. The technique is described as a hybrid method that combines the qualitative and quantitative approach in a single research exercise (Sekayi & Kennedy, 2017).

10.3.1   Targeted sample

The two case studies used for this paper are, firstly, the development of KPIs for the measurement of the performance of the FM unit in an HE institution. In this research, the participants or panel of experts were drawn from the strategic and tactical levels of leadership among the client – dean of faculties, consumer or end-users – head of academic departments and the service providers – FM unit. To qualify for the exercise, each participant must have spent at least one year in the office or five years in their respective departments. A population of 78 participants was targeted, but 39 effectively responded, as shown in Table 10.1. They were also invited to participate in the second phase of the exercise aimed at developing the

Table 10.1 Response to first and second phase of data collection

Classification	Response to 1st phase						2nd phase		
	Target	Silent	Decline	Interviewed	Will you participate in 2nd phase?		*1*	*2*	*3*
					Yes	No			
Academic strategic: Deans	9	2	–	7	4	3	2	2	2
Academic tactical: HODs	57	25	12	20	16	4	9	7	7
FM Strategic	4	–	–	4	3	1	2	2	2
FM Tactical	8	–	–	8	8	–	4	4	3
Total	**78**	**27**	**12**	**39**	**31**	**8**	**17**	**15**	**14**

KPIs, using the Delphi technique. Initially, 31 persons accepted to participate but only 17, 15 and 14 eventually participated in the three rounds of data collection.

The second study focused on the examination of the factors responsible for the delays in the execution of a construction project where there is evidence of adequate funding. Similarly, the participants were the Directors of Physical Planning (DPP) and the Directors of Works (DOW), who are professionals from the engineering and the built environment professions (engineers, builders, quantity surveyors and architects). Each person should have at least ten years, post-professional registration experience and must have spent at least five years in office as a DPP or DOW. Five universities were randomly selected, with two participants representing each university. The ten participants responded in round 1 but 6 persons in rounds 2 and 3, as shown in Table 10.2.

10.3.2 Procedure

In the first research, the "modified classic" Delphi approach was used for data collection (Franklin & Hart, 2007). In this approach, a generic list of 112 KPIs, divided into seven categories, was circulated to the participants in the first round. They were requested to score the KPIs on a Likert scale of 1–5, with one being the lowest rating and five the highest. It was agreed that only the items that scored 3.0 and above will be escalated to the next round. At the end of the first and second

Table 10.2 Participants in the Delphi exercise

Participants	Round 1		Round 2		Round 3	
	Yes	No	Yes	No	Yes	No
DPP	5	-	3	2	3	-
DOW	5	-	3	2	3	-
Total	**10**	-	**6**	**4**	**6**	-

Table 10.3 List of KPIs round 1

S/No	Description	Score
	Capital development: Process	
1	Effective representation of project briefing into developed asset	3.33
2	**Reduced dispute and litigation**	**2.33**
3	Incorporate end-users into project execution team	3.4
4	Conduct end-user's orientation into new facilities	3.27
5	The new facility should increase the positive reputation of end-users	3.53
6	**Produce effective "as-built" documents of the completed facility on handover**	**2.8**
7	Prompt correction of faults	3.60
8	Enhance new technological capability	3.0
9	**Contributing to the effectiveness of other projects of end-users**	**2.73**
10	**Functional in operation to reduce dependence on outside source**	**2.60**

rounds, 104 and 94 KPIs respectively achieved the benchmark. In the third round, participants did not change their opinions. Therefore, the 94 KPIs were accepted, arranged in their order of priority and communicated to all participants. The summary of the resulting KPIs is shown in Table 10.3.

Similarly, in the second study, which was a pilot study, the "classic Delphi" approach was adopted (Franklin & Hart, 2007). In this approach, the first round was a qualitative response where participants generated the proposed reasons for the delay in the execution of a construction project. Each participant was requested to suggest between three and five reasons for the delay. In all, they suggested 30 reasons. After analysis, observing a similar benchmark as in the first case study, the 30 reasons were reduced to eight and six factors in rounds 2 and 3 respectively as is seen in Tables 10.7 and 10.8. After round 3, the response was stable and the suggested reasons for the delays were classified as external (having to do with the finding agency) and internal, bordering on the internal operations of the respective universities (See Table 10.9).

10.3.3 Methods of analysis

As a result of the small size of the research sample in both cases, the simple mathematical average was used for the analysis, instead of the rigorous statistical analysis. The scale of measure was the Likert scale of 1–5. Only items which scored 3.0 and above were escalated to next round and also used as the benchmark for consensus. The details of how the Delphi technique was used in both the research and the research findings are discussed in the next section.

10.4 Results and findings

In the first case study the generic list of 112 KPIs was divided into seven sub-headings (some with other sub-sub-headings). The section on capital development: Process, is used to illustrate the procedure adopted. Table 10.3

Table 10.4 List of KPIs for rounds 2 and 3

S/No	Description
	Capital development: Process
1	Effective representation of project briefing into developed asset
2	Incorporate end-users into project execution team
3	Conduct end-user's orientation into new facilities
4	The new facility should increase the positive reputation of end-users
5	Prompt correction of faults
6	Enhance new technological capability

shows the ten items on the generic list of KPIs circulated to the participants in round 1. After the analysis of the submissions from the participants, only six items satisfied the requirements of the benchmark of 3.0 and above. The other items marked in red, below the benchmark, were not escalated to the next round.

The analysis of round 1 was sent to all the participants, showing the items that met the benchmark and those that did not. They were requested to identify any item that did not meet the benchmark, but what they felt should be considered and included in the next round, using a different colour code. No participant suggested any item from those that did not meet the benchmark. The six items that met the requirements of the benchmark were circulated to participants in rounds 2 and 3, as shown in Table 10.4. After each analysis, the six items were retained, an indication that the process had attained equilibrium or consensus.

However, the score for each item in the third round suggested a re-arrangement of the list in their order of priority. The six KPIs were arranged accordingly, as shown in Table 10.5. This table represents the consensus of opinion of all the participants.

Finally, the priority list was classified as shown in Table 10.6, which represents the findings of this research (part of the 94 KPIs) that was communicated to all the participants at the end of the exercise.

Table 10.5 Priority list of KPIs

S/No	Description	Score
	Capital development: Process	
1	Prompt correction of faults	4.29
2	The new facility should increase the positive reputation of end-users	4.21
3	Conduct end-user's orientation into new facilities	3.64
4	Enhance new technological capability	3.64
5	Incorporate end-users into project execution team	3.71
6	Effective representation of project briefing into developed asset	3.57

Table 10.6 Priority list and classification of KPIs

		Description, classification and rating		
S/No	Category	High priority (4.5–5.0)	Medium priority (4.0–4.49)	Low priority (3.5–3.99)
1	**Capital development**			
a	Process		Prompt correction of faults	Incorporate end-users into project execution team
			The new facility should increase the positive reputation of end-users	Conduct end-user's orientation into new facilities
				Enhance new technological capability
				Effective representation of project briefing into developed asset

Similarly, in the second case study the first round was a qualitative response to an open-ended question on the subject of the research which reads as follows:

What are the factors responsible for delays in the execution of Tetfund projects on schedule?

Each participant was requested to identify between three and five factors they considered to be responsible for the delays in the execution of construction project funded by this particular agency.

One of the respondents suggested the following five reasons:

1   Delay in receiving letter of allocation
2   Delay in receiving AIP
3   Delay in mandatory monitoring, evaluation and project inspection
4   Delay in receiving first tranche
5   Delay in receiving second tranche

After the collation of responses from all ten respondents, 30 reasons were suggested. In round 2, participants were requested to score the 30 items on a scale of 1–5, similar to the first case study, adopting 3.0 as the benchmark. The result, after the second round, reduced the 30 items to eight, as shown in Table 10.7.

In the third round, two more items were eliminated, because, they scored below the benchmark, as shown in Table 10.8.

Literature suggests that consensus can be reached if participants agree on 51–80% of the items on the list for interaction. In this regard, participants attained 75% consensus that is agreement on six out of the eight items in this round. These were accepted as the solution to the research question on the causes of delay in the execution of construction projects funded by the agency under reference. The findings were arranged as external and internal factors, as shown in Table 10.9.

Table 10.7 Result of the analysis of round 2

S/No	Suggested reasons	Analysis of round 2
1	Delay in receiving letter of allocation	**2.5**
2	Delay in receiving AIP	3.25
3	Delay in mandatory monitoring, evaluation and project inspection	3.0
4	Delay in receiving first tranche	4.75
5	Delay in receiving second tranche	3.5
6	Economic factor of the contractor	**2.25**
7	Ill-conceived projects	**1.5**
8	Delay in harmony of payment certificates	**2.0**
9	Contract awarded to incompetent contractors	4.0
10	Inability to meet conditions of release of funds by beneficiaries on time	**2.75**
11	Frequent change in design	**1.5**
12	Hostility of host community	**1.25**
13	Late honouring of certificates by client	**2.0**
14	Force majeure	**1.25**
15	Incomplete architect instruction	**1.75**
16	Contractors not receiving instruction/drawing/other details on time	3.25
17	Requesting gratifications from contractors	**1.5**
18	Incompetent technical in-house staff	**2.0**
19	Using inferior materials	**1.75**
20	Bad workmanship requiring reworks	**2.25**
22	Non-completion of tranches before release of another by the institutions	**1.0**
23	Non-submission of observation by the institution when requested by Tetfund	**1.5**
24	Delay in calling Tetfund for inspection in order to access the next tranche	**2.25**
25	Delay may be by the contractor	4.25
26	Wrong contractor selection method	**1.75**
27	Lack of flexibility in fund utilization (market realities)	**2.25**
28	Non-completion of projects affects accessing future funds	3.25
29	Contractor always holds client to ransom	**1.5**
30	Time taken to obtain approvals always attract fluctuation of price	**2.25**

Table 10.8 Analysis of round 3

S/No	Suggested reasons after first analysis	Score
1	**Delay in receiving AIP**	2.75
2	Delay in mandatory monitoring, evaluation and project inspection	3.0
3	Delay in receiving first tranche	3.0
4	Delay in receiving second tranche	3.5
5	Contract awarded to incompetent contractors	3.0
6	**Contractors not receiving instruction/drawing/other details on time**	2.5
7	Delay may be by the contractor	4.0
8	Non-completion of projects affects accessing future funds	3.75

Table 10.9 The research findings

S/No	External factors	Score
1	Delay in receiving first tranche	3.0
2	Delay in receiving second tranche	3.5
3	Delay in mandatory monitoring, evaluation and project inspection	3.0
	Internal factors	
4	Delay may be by the contractor	4.0
5	Non-completion of projects affects accessing future funds	3.75
6	Contract awarded to incompetent contractors	3.0

The next section of this paper will discuss the validity and reliability of this tool in conducting credible research.

10.5 Validity and reliability

There are three inter-related components in the Delphi process as a research tool that should be given detailed consideration when discussing the question of the validity and reliability of the technique. The components are: the quality of the participants, the coordinator and anonymity of participants' contributions as well as the method of achieving consensus. These are essential factors that influence the credibility of the research results.

10.5.1 The quality of participants

The role of participants in a Delphi exercise is critical and influences the success of the research effort. As a tool for consensus building, the strength of the Delphi technique stems from the concept or theory that "several people are less likely to arrive at a wrong decision than a single individual" (Hasson et al., 2000, p. 1013). The "several people" here refers to experts, informed individuals or professionals in the field of study. The method of selecting the participants closely follows the "purposive or criterion sampling" rather than the random sampling methods, used in general surveys (Hasson et al., 2000). This is because the participants are selected to apply their knowledge or expertise to address the subject of the research. The participants so chosen within this elastic continuum of the inclusive population, without restrictions on geographical locations, are broad-minded individuals with professional or experiential knowledge of the subject of the research. They should be willing to dedicate themselves to the repeated process of the Delphi exercise (Donohoe & Needham, 2009). The participants may or may not know one another. Nevertheless, they are persons of high integrity, with independent minds and whose opinions can be trusted. It is imperative, therefore, to adopt suitable pre-qualification criteria in selecting participants in line with the research objectives.

The participants for the case study were drawn from within the same university in South Africa. The objective of the exercise was to develop suitable KPIs for measuring the performance of the service provider FM unit. The function of the FM unit is to develop, operate and maintain the necessary support facilities and services suitable for the performance of the core functions of teaching, learning and research. The composition of the panel of participants (as shown in Table 10.2) was seven of the nine deans of faculties that participated in the first phase of data collection. Four indicated an interest in participating in the second phase, but due to other commitments, only two followed through to the end of the process. Invitations were extended to 57 heads of academic departments, but 20 participated in phase one, nine, seven and seven in the three rounds of phase two. The deans are highly knowledgeable persons who understand and effectively participate in the development of the content, processes and expected outputs of each academic curriculum as well as the quality of facilities and facility services required for the effective delivery of the process. The leaders at the tactical levels, also known as end-users, are the heads of academic departments. These are the leaders who coordinate the day-to-day activities which facilitate teaching, learning and research within a suitable academic environment. They interact with the service provider in order to maintain the quality and functional state of the facilities and facility services suitable for the performance of their core functions. Similarly, the strategic and tactical leaders among the service provider (FM unit) are professionals from the engineering and the built environment professions. They translate clients' briefs into the development of new infrastructure or remodel existing ones to suit particular academic programmes. They coordinate the operation and maintenance of existing facilities and facilities services throughout their life cycle in order to maintain an academic environment suitable for the performance of the core functions of teaching, learning and research.

Similarly, the participants in the second case study are the strategic leaders responsible for the development of infrastructure and services (DPP) and those responsible for the coordination of operation and maintenance (DOW), in their respective universities, as shown in Table 10.3. The participants were drawn from five different universities, in the same country, Nigeria. These individuals are certified professionals from the engineering and the built environment professions with years of experience. They are experts in their professions who were considered suitable to address the research question on the causes of delays in the execution of construction projects funded by the particular agency of government.

Literature suggests that the size of a Delphi panel may be as small as three members and as large as possible (Mullen, 2003; Grisham, 2009; Kezar & Maxey, 2016). However, if the composition of participants is homogeneous, a small number of panelists suffice. The subjects that were addressed in the two case studies were issues of common interest to the participants and the participants from the target population were homogenous. This has accounted for the size of participants, without compromise on the quality and reliability of the research outcome.

Adopting appropriate pre-qualification criteria allows the research coordinator to recruit suitable participants from the pool of prospects available, without restriction of geographical location. This is in tandem with best practice gleaned from literature.

The research objective of Grisham (2009) was to test the hypothesis that "there are universal attributes for cross-cultural leadership that are effective regardless of culture" (Grisham, 2009, p. 119). In this study, the boundary conditions set for participants were that the prospective participants should have "at least 20 years of practical experience working in an international or multicultural environment, in any industry or a person that has an advanced degree in leadership or cross-cultural studies with over 20 years of research, teaching, and publication experience; or a combination of the two" (Grisham, 2009, p. 121). Within this envelop, he raised a panel of 25 experts from different regions of the world to include: Eastern Europe, Nordic Europe, Germanic Europe, Latin Europe, Latin America, Confucian Asia, Southern Asia, Anglo Sub-Saharan Africa and the Middle East (Grisham, 2009). Specifically, considering the engineering and the built environment research fields, Hallowell and Gambatese (2010) suggested a broad-based set of qualifications; an excerpt is shown in Table 10.10. In their opinion, prospective experts must satisfy at least four of the following criteria related to the research topic and score a minimum total of 11 points.

In a similar exercise, Musonda and Pretorius (2015) were interested in the study of health and safety issues in the construction industry in South Africa. They invited experts knowledgeable in the field from those within the contracting, consulting industries and academics. They further narrowed their search to

Table 10.10 Qualification for panel of experts (after Hallowell & Gambatese, 2010, p. 4)

Characteristic	Minimum requirement (Achievement or experience)	Point (each)
Qualifying panelists as experts	Professional registration such as Professional Engineer (PE), Licensed Architect (AIA), Certified Safety Professional (CSP), Associated Risk Manager (ARM)	3
	At least 5 years of professional experience in the construction industry	1
	Invited to present papers at a conference	0.5
	Member of a nationally recognized committee	1
	Chair of a nationally recognized committee	3
	Primary or secondary writer of at least three peer-reviewed journal articles	2
	Faculty member at an accredited institution of higher learning	3
	Writer/editor of a book	4
	Writer of a book chapter	2
	Advanced degree in the field of engineering, CEM or other related fields	
	BS	4
	MS	2
	PhD	4

individuals who had frequently appeared as authors or keynote speakers in the conference proceedings of the CIB WO99 from 2005 to 2009. The resulting panel of eleven (11) represented persons as follows: two members from South Africa, three each from the United States of America (USA) and the United Kingdom (UK), and one each from Singapore, Hong Kong and Sweden. Three of the panelists were employed by contracting organizations, two by consulting firms, and six by universities (Musonda & Pretorius, 2015).

10.5.2 The coordinator and anonymity of participants' contributions

The anchor person in a Delphi exercise is severally referred to as the researcher, facilitator or simply called the coordinator. The coordinator pilots the activities in the exercise from the research design, selection of the participants, managing the process to writing the final report (Kezar & Maxey, 2016; Ogbeifun, *et al*, 2016). Some questions the coordinator should address up front include panel design, question clarity, managing the process, managing attrition rate and outliers as well as the smooth progression of the exercise and final reporting (Hasson *et al.*, 2000; Donohoe & Needham, 2009; Geist, 2010). The coordinator serves as an unbiased umpire, filters the submissions of the participants and records them with appropriate codes so that no participant can be traced to their contribution. After the analysis of each submission, the coordinator recycles the feedback to participants with instructions on the next line of action until consensus is achieved.

Unlike the general survey, where participants respond to the survey questions once, in the Delphi approach participants respond several times until equilibrium is achieved. In the process, participants change their views from one round to another. When participants change their opinions, especially when these changes are unique to the individual participants, it demonstrates the individual's response to the collective wisdom of the group. Integral to the Delphi process is the expectation that participants will change their opinion during the exercise. This is because participants have several opportunities to interact with the same subject over a lengthy period, as well as to reflect on the submissions of other participants (Makkonen *et al.*, 2016).

As shown in Table 10.11, the second case study for this paper a Director of Physical Planning (DPP1) and a Director of Works (DOW1) as well as DPP2 and DOW 2 are from two separate universities. Although the participants knew one another, their responses to the research questions are significantly different. They changed their opinions from one round of the exercise to another (rounds 2 and 3) without any indication of complicity. The final conclusion is that it is possible to achieve consensus of opinion in a Delphi exercise through the anonymity of participants' contributions, even when the participants knew one other (Ogbeifun *et al.*, 2017). This is one of the strong defences for the use of the Delphi technique. It can be achieved through painstaking selection of appropriate participants, the skills of the research coordinator and transparent communications between the rounds as well as in the final report.

Table 10.11 Raw data of participants from the same institution

S/No	Suggested reasons	DPP1	DOW1	DPP2	DOW2
			Rating		
The eight items that met the benchmark in round 2					
1	Delay in receiving AIP	2	4	2	5
2	Delay in mandatory monitoring, evaluation and project inspection	3	3	3	3
3	Delay in receiving first tranche	5	4	5	5
4	Delay in receiving second tranche	3	3	3	5
5	Contract awarded to incompetent contractors	4	3	4	5
6	Contractors not receiving instruction/ drawing/other details on time	4	2	4	3
7	Delay may be by the contractor	5	3	5	4
8	Non completion of projects affecting assessment of future funds	2	4	2	5
	Submission for round 3				
1	**Delay in receiving AIP**	3	3	4	1
2	Delay in mandatory monitoring, evaluation and project inspection	4	2	4	2
3	Delay in receiving first tranche	2	2	4	4
4	Delay in receiving second tranche	2	4	4	4
5	Contract awarded to incompetent contractors	4	3	4	1
6	**Contractors not receiving instruction/ drawing/other details on time**	3	3	3	1
7	Delay may be by the contractor	4	3	4	5
8	Non-completion of projects affecting the assessment of future funds	5	5	3	2

10.5.3 *The method(s) of achieving consensus*

The third critical component of the Delphi technique is the method of achieving consensus. The procedure in a typical Delphi exercise is to circulate the information on the proposed solutions to the research question to the participants, collate their responses and after analysis, to re-circulate the feedback to the panel members. The process continues until consensus is achieved. There are no firm rules regarding the number of rounds in the Delphi exercise, one or many rounds of information gathering suffice. Consensus or convergence of opinion can be achieved when the participants are no longer modifying their earlier decisions (Franklin & Hart, 2007; Adnan & Daud, 2010) or the participants are in agreement in between 51% and 80% of the suggested solutions (Hasson *et al.*, 2000). The method of determining convergence of opinion should be spelt out at the beginning of the exercise, applied and communicated to all participants through the different rounds and in the final report (Day & Bobeva, 2005). Where there are fewer than 20 participants, the simple mathematical mean can be used for analysis; otherwise the standard statistical analysis should be adopted.

In the example given in Table 10.4, ten generic KPIs were circulated to participants in round 1. After analysis, six items met the requirements of the set benchmark and were escalated to the second round. In the second and third rounds (Table 10.5), these six items were retained. That is, participants did not change their opinions on these items. Therefore, equilibrium or consensus had been achieved, as shown in Table 10.6. However, in the second case study, the ten individuals who participated in round 1 of the exercise contributed 30 potential solutions to the research question, as shown in Table 10.8. These 30 ideas were circulated to the participants in round 2. After analysis of their responses only eight items met the requirements of the benchmark and were escalated to round 2, as shown in Table 10.9. The result of the analysis of the responses in round 2 was circulated to all participants for further interaction in round 3. After analysis of the responses in round 3, two other items (25%) did not meet the benchmark. Therefore, the participants attained 75% consensus of opinion and the exercise was terminated. The solution to the research question was arranged and classified as external and internal factors, as shown in Table 10.10.

The credibility of the two research reports was that the participants accepted the results and were willing to work with them in order to find solutions to problems affecting the operations of their respective institutions. Researchers suggest that the Delphi technique can be used as a stand-alone tool for credible research or can be used in conjunction with other research tools. If those who participated in the Delphi exercise are part of the panel of experts, the results can be accepted and put to use. However, if the exercise is conducted and the result is to be implemented by other interested parties, it is suggested that the outcome should be subjected to other research methods for fine tuning; such other methods could be focus group sessions or another Delphi exercise.

10.6 Conclusions

The need to conduct quality research has been a continuous concern to researchers in all fields, including the built environment professions. This desire challenges researchers to pay close attention to the data used for any research exercise, the reliability of the source(s) of the data and the tool(s) used for data collection. In this regard, several research approaches are being developed in order to improve the reliability and validity of research outcomes. To this effect, the Delphi technique, which is a hybrid method that combines the qualitative and quantitative approach in a single research exercise, is increasingly being used as a reliable research tool in engineering and the built environment studies. The adoption of this approach stems from the concept that many like-minded people, participating in a research project are less likely to arrive at a wrong decision than a single individual. In order to achieve this objective, the participants in a typical Delphi exercise are selected "purposively". These participants are knowledgeable persons or professionals in the area of the research, sourced from a wide continuum without the barriers of geographical location. The Delphi process is repetitive and can be slow. It, therefore, requires commitment to go through repetitive cycles of the exercise, thus necessitating the recruitment of a large number of participants at the beginning of the exercise in order to ameliorate the negative effects of attrition. The strength of the technique revolves around the quality of the participants (usually selected through defined pre-qualification criteria), the anonymity of participants' contributions and results being refined through successive rounds of data collection. Equilibrium or consensus is reached through the continuous recycling of results among participants within a few or many rounds of successive iterations. The research coordinator manages the controlled feedback process, which allows participants to view their individual submissions in the light of the perspective of whole group. In most cases, participants adjust their opinions from one round to another without coercion or complicity.

This technique was used in the two case studies discussed in this paper. By observing the pre-qualification principles integral in the Delphi technique, suitable participants were recruited for the two exercises. In the first case study, the academics at the strategic and tactical levels of leadership were satisfied that the developed KPIs will assist the services provider (FM unit) in the performance of their functions, which will in turn facilitate the performance of the core functions of the academics. On the part of the service provider, the FM unit, they were enthusiastic to use the developed KPIs, seeing the tool as a panacea to their efforts in developing, operating and maintaining functional facilities suitable for the performance of the core functions of teaching, learning and research within a productive academic environment.

Similarly, the DPPs and DOWs in the second case study were encouraged by the results from the Delphi exercise and the classification of the resulting factors into external and internal factors. The result revealed that the internal operatives of the respective institutions have their role to play in addressing the causes of delay in the execution of the construction projects. While the external factors identified provides the DPP and DOW with useful information on how to objectively engage with the funding agency.

The results from the cited case studies have demonstrated that the Delphi technique is suitable for the conduct of reliable and valid research within the built environment. The adjustments of participants' opinions from one round to the other is central to the refining process of the results from a Delphi exercise; a considerable advantage over the outcome of a typical general survey where decisions are taken based on the once-off contribution from participants.

References

Adnan, Y.M. and Daud, M.N. (2010), Factors influencing office building occupation decision by tenants in Kuala Lumpur City Centre – A Delphi study, *Journal of Design and Built Environment*, Vol. 6, June, pp. 63–82.

Agumba, J.N., Haupt, T.C. and Pretorius, J.H.C. (2014), Important health and safety performance improvement indicators for small & medium construction enterprises in South Africa: Eliciting expert opinion using the Delphi technique, *Journal of the Ergonomics Society of South Africa*, Vol. 26, No. 2, pp. 3–22.

Alaloul, W.S., Liew, M.S., Wan, Z. and Noor, A. (2015), Delphi technique procedures: A new perspective in construction management research, *Applied Mechanics & Materials*, Vol. 802, pp. 661–667.

Day, J. and Bobeva, M. (2005), A generic toolkit for the successful management of Delphi studies, *The Electronic Journal of Business Research Methodology*, Vol. 3, pp. 103–116.

Donohoe, H.M. and Needham, R.D. (2009), Moving best practice forward: Delphi characteristics, advantages, potential problems, and solutions, *International Journal of Tourism Research*, Vol. 11, pp. 415–437.

Franklin, K.K.J. and Hart, K. (2007), Idea generation and exploration: Benefits and limitations of the Policy Delphi research method, *Innovative Higher Education*, Vol. 31, pp. 237–246.

Gerring, J. (2004). What is a case study and what is it good for? *American Political Science Review*, Vol. 98, pp. 341–354.

Geist, M.R. (2010), Using the Delphi method to engage stakeholders: A comparison of two studies, *Evaluation and Programme Planning*, Vol. 33, pp. 147–154.

Green, J. and Thorogood, N. (2009) "*Qualitative methods for health research*", 2nd Ed, London: SAGE Publication, Ltd.

Grisham, T. (2009), The Delphi technique: A method for testing complex and multifaceted topics, *International Journal for Managing Projects in Business*, Vol. 2, pp. 112–130.

Habibi, A., Sarafrazi, A. and Izadyar, S. (2014), Delphi technique theoretical framework in qualitative research, *The International Journal of Engineering and Science*, Vol. 3, No. 4, pp. 8–13.

Hallowell, M.R. and Gambatese, J.A. (2010), Qualitative research: Application of the Delphi method to CEM research, *Journal of Construction Engineering and Management*, Vol. 136, pp. 1–9.

Hallowell, M.R., Hinze, J.W. and Baud, K.C. (2013), Proactive construction safety control: Measuring, monitoring, and responding to safety leading indicators, *Journal of Construction Engineering and Management*, Vol. 139, No. 10. Available online at: https://doi.org/10.1061/(ASCE)CO.1943-7862.0000730, Accessed 10 June 2020

Hasson, F., Keeney, S. and McKenna, H. (2000), Research guidelines for the Delphi survey, *Journal of Advance Nursing*, Vol. 32, p. 1008–1015.

Hasson, F. and Keeney, S. (2011), Enhancing rigour in the Delphi technique research, *Technological Forecasting & Social Change*, Vol. 78, pp. 1695–1704.

Holliday, A.R. (2007), *Doing and writing qualitative research*, 2nd Ed., London: Sage.

Kermanshachi, S., Dao, B., Shane, J. and Anderson, S. (2016), An empirical study into identifying project complexity management strategies, *Procedia Engineering*, Vol. 145, pp. 603–610.

Kezar, A. and Maxey, D. (2016), The Delphi technique: An untapped approach of participatory research, *International Journal of Social Research Methodology*, Vol. 19, No. 2, pp. 143–160.

Lateef, O.A., Khmidi, M.F. and Idrus, A., (2010) Building maintenance management in a Malaysian University campus: A case study, *Australian Journal of Construction Economics and Building*, UTS e PRESS, vol. 10, No. 1/2, 76–89.

Makkonen, M., Hujala, T. and Uusivuori, J. (2016), Policy expert's propensity to change their opinion along Delphi rounds, *Technological Forecasting & Social Change*, Vol. 109, August, pp. 61–68.

Mullen, P.M. (2003), Delphi myths and reality, *Journal of Health Organisation and Management*, Vol. 17, pp. 37–52.

Musonda, I. and Pretorius, J.H.C. (2015), Effectiveness of economic incentives on clients' participation in health and safety programmes, *Journal of the South African Institution of Civil Engineering*, Vol. 157, Available online at: http://www.scielo.org.za/scielo.php?pid=S1021-20192015000200001&script=sci_arttext Accessed 2 November 2016.

Ogbeifun, E., Agwa-Ejon, J., Mbohwa, C. and Pretorius, J.H.C. (2016), The Delphi technique: A credible research methodology, Proceedings of the *Sixth International Conference on Industrial Engineering and Operations Management*, ISBN: 978-0-9855497-4-9, ISSN: 2169-8767, Kuala Lumpur, Malaysia, March 8-10, 2016, pp. 2004–2009 Available at: http://ieomsociety.org/ieom_2016/pdfs/589.pdf Accessed 6 February 2019.

Ogbeifun, E., Mbohwa, C. and Pretorius, J.H.C. (2017), Achieving consensus devoid of complicity: Adopting the Delphi technique, *International Journal of Productivity and Performance Management*, Vol. 66, No. 6, pp. 766–779.

Sekayi, D. and Kennedy, A. (2017), Qualitative Delphi method: A four round process with a worked example, *The Qualitative Report*, Vol. 22, No. 10, pp. 2755–2763.

Yin, R.K. (2014), *Case study research- design and methods*, 5th Ed., Singapore: Sage.

Zainal, Z. (2007), Case study as a research method, *Jurnal Kemanusiaanbil.* Vol. 9, No. 1, pp. 1–6.

Index

Note: Italicized and bold page numbers refer to figures and tables.

For Product Safety Concerns and Information please contact our EU
representative GPSR@taylorandfrancis.com
Taylor & Francis Verlag GmbH, Kaufingerstraße 24, 80331 München, Germany

www.ingramcontent.com/pod-product-compliance
Lightning Source LLC
Chambersburg PA
CBHW070730220326
41598CB00024BA/3381